THE NEW WILD

WHY INVASIVE
SPECIES WILL BE
NATURE'S SALVATION

FRED PEARCE

Published in the UK in 2015 by
Icon Books Ltd, Omnibus Business Centre,
39–41 North Road, London N7 9DP
email: info@iconbooks.com
www.iconbooks.com

Sold in the UK, Europe and Asia
by Faber & Faber Ltd, Bloomsbury House,
74–77 Great Russell Street,
London WC1B 3DA or their agents

Distributed in the UK, Europe and Asia
by TBS Ltd, TBS Distribution Centre, Colchester Road,
Frating Green, Colchester CO7 7DW

Distributed in Australia and New Zealand
by Allen & Unwin Pty Ltd,
PO Box 8500, 83 Alexander Street,
Crows Nest, NSW 2065

Distributed in South Africa by
Jonathan Ball, Office B4, The District,
41 Sir Lowry Road, Woodstock 7925

Distributed in India by Penguin Books India,
7th Floor, Infinity Tower – C, DLF Cyber City,
Gurgaon 122002, Haryana

ISBN: 978-184831-834-2

Prin [obscured] c

THE NEW WILD

Contents

About the author

Fred Pearce has been environment and development consultant at *New Scientist* magazine since 1992, reporting from 85 countries. He has written fourteen previous books, including *When the Rivers Run Dry* and *The Land Grabbers*, which have been translated into 23 languages. He also writes for the *Guardian* and is a regular broadcaster on radio and TV. He won a lifetime achievement award for his journalism from the Association of British Science Writers in 2011, and was voted UK Environment Journalist of the Year in 2001.

Acknowledgements

The roll call of people who helped in this book goes back through most of my 35 years writing on science and the environment. Many are named and quoted in the text. But some deserve special mention for helping me on my travels. On Ascension Island, Stedson Stroud was a marvellous mentor and guide; Paul Lister invited me to Scotland's Alladale Wilderness Reserve; Geoffrey Howard hosted me in East Africa; Nicola Divine McClain lured me to Montana; Peter Shaw took me to see his Essex ashpits and James Fraser helped me in Liberia. My thirst for the contrary has also been nurtured at successive meetings of the Breakthrough Institute in California, where several of the interviews here took place. So, my thanks to Michael Shellenberger and Ted Nordhaus for inviting me.

Many editors at *New Scientist* have helped fund my inquiries over the years. Kate Douglas sent me to Lake Victoria and David Concar to Sarawak. Thanks also to Jeremy Webb, Bill O'Neill, Michael Bond, Sumit Paul-Choudhury, Rowan Hooper and Graham Lawton. Other commissioners whose enthusiasms have helped develop my ideas on the new wild over the years include Brian Leith for Granada TV, Bruce Stutz at Audubon (who sent me to investigate the Mediterranean's 'killer algae'), Susanna Wadeson at Transworld, and Roger Cohn at Yale e360.

Many people gave me interviews while I worked on this book. They include Chris Thomas, Joseph Mascaro, Nicola Weber, Paul Robbins, Andrew Cohen, Melissa Leach, Erle Ellis, Stephen Pyne, Borgthor Magnusson, Jay Stachowicz, Ariel Lugo, Rick Shine, Peter Kareiva, Douglas Sheil, Stewart Brand, Joe DiTomaso, Sean Hathaway, Dick Shaw, Matt Shardlow,

Jim Dickson, David Wilkinson, Susan Schwartz, Ken Collins, Jeffrey Sayer and William Laurance. Many others responded at length to my emails. Still more, some quoted and some forgotten, did the same for previous journalism that spun into this book. Thanks to them all.

On language

I have used a lot of words in this book to describe species moving to new places. Most frequently I call them aliens, not in a pejorative sense, but to describe their status. More for variety than anything else, I also call them vagabonds, invaders, carpetbaggers, migrants, interlopers, non-natives, incomers, and no doubt more. Some aliens were taken by humans deliberately or by accident; others arrived by more traditional means. I save the adjective 'introduced' for those taken by humans. For some ecologists, 'invasive' has a particular meaning. It applies to those aliens making waves and taking territory. Equally, when species settle down and appear at home, they will be termed by some 'naturalized'. I generally follow both terms. But please judge the species I describe by what they do or don't do, and not by the label attached to them.

Some people like the Latin names of plants and animals, and sometimes they are necessary to clear up confusion. But they do get in the way. So I have included Latin names in the main text for the main species and taxa when I first mention them, but not for bit-part players. For those, where the context allows me to be specific about individual species, and where I think readers may be interested, I have posted an appendix containing some common names and their Latin versions.

Acronyms

CBD	Convention on Biological Diversity
DDT	Dichlorodiphenyltrichloroethane, a pesticide
FAO	Food and Agriculture Organization of the United Nations

GCHQ	Government Communications Headquarters (UK)
IMO	International Maritime Organization
IPCC	Intergovernmental Panel on Climate Change
IUCN	International Union for Conservation of Nature
JK	Japanese knotweed
MEA	Millennium Ecosystem Assessment
MIT	Massachusetts Institute of Technology
NSA	National Security Agency (US)
RSPB	Royal Society for the Protection of Birds
TNC	The Nature Conservancy (US)
UC	University of California
UNEP	United Nations Environment Programme
UNESCO	United Nations Educational, Scientific and Cultural Organization
USGS	US Geological Survey
WRI	World Resources Institute
WWF	World Wildlife Fund

Introduction

Nature in a World of Humans

Rogue rats, predatory jellyfish, suffocating super-weeds, wild boar, snakehead fish wriggling across the land – alien species are taking over. Nature's vagabonds, ruffians and carpetbaggers are headed for an ecosystem near you. These biological adventurers are travelling the world in ever greater numbers, hitchhiking in our hand luggage, hidden in cargo holds, stuck to the bottom of ships and migrating to keep up with climate change. Our modern, human-dominated world of globalized trade and messed up ecosystems is giving footloose species many more chances to cruise the planet and set up home in distant lands. Some run riot, massacring local species, trashing their new habitats and spreading diseases.

We all like a simple story with good guys and bad guys. And aliens always make easy enemies. So the threat of foreign species invading fragile environments and causing ecological mayhem gets our attention. Conservationists have for half a century been battling to hold back the tide of aliens. They call them the second biggest threat to nature, after habitat loss. Their concern is laudable. They want to protect native species and the ecosystems they inhabit. But do we fear these ecological outsiders too much? Is our fear usually little more than green xenophobia? Most of us are appalled when foreigners are treated as somehow intrinsically dangerous. Yet the orthodoxy in conservation is to demonize foreign species in just that way. Native is good, and alien is bad. But is this simple formula true? Or might we need the go-getter, can-do aliens? In fact, might their success be a sign of nature's resilience in the face of the considerable damage humans have done to the planet?

I am an environment journalist. Even to ask such questions gets me treated in some circles as a conservation heretic. I have met incredulity and hostility in equal measure. To be clear, I am not accusing environmentalists of being closet xenophobes or misanthropes, still less racists. But I have found that I am far from alone in my concern that we have bought into some dangerous mythology about how nature works.

I am not questioning the motives – to strengthen nature – but the means. Many ecologists who actually study nature told me that they felt conservationists were, with the best of intentions, getting the aliens wrong. And worse, that their efforts to keep out all foreign invaders of ecosystems might often be counter-productive, weakening nature rather than strengthening it. I discovered that there is a scientific backlash going on against the simple formula that natives are good and aliens bad. The purpose of this book is to explore that new thinking, and to ask what it should mean for conservation.

My conclusion is that mainstream conservationists are right that we need a re-wilding of the Earth, but wrong if they imagine that we can achieve that by going backwards. We need a New Wild – hence the title of this book. But the new wild will be very different to the old wild. We have changed our planet too much, and nature never goes backwards. Nature's resilience is increasingly expressed in the strength and colonizing abilities of alien species. They are often the new natives. And in the new wild, we need to stand back and applaud.

There are horror stories of alien takeovers, of course. Most of them happen on small, remote islands with only a few native species, where carnivorous rats, cats and others hop off ships and cause mayhem. But elsewhere, most of the time, the tens of thousands of introduced species usually either swiftly die out or settle down and become model eco-citizens, pollinating crops, spreading seeds, controlling predators, and providing food and habitat for native species. They rarely eliminate natives. Rather than reducing biodiversity, the novel new worlds that result are usually richer in species than what went before. Even the terror suspects of conservation, such as zebra mussels and tamarisk, Japanese knotweed and water hyacinth, often have a good side we rarely hear about.

After going on the trail of alien species across six continents, my conclusion is that their demonization says more about us and our fears of change than about them and their behaviour. Some ardent wildlife lovers show a dark side when it comes to aliens. I sometimes think the more ardent they are, the more likely they are to be rabid about alien species. Understandable love of the local, the native and the familiar – of an imagined pristine environment before humans showed up – too often becomes fear and hatred of the foreign and the unfamiliar.

This hostility is generally justified by outdated and ill-founded ideas about how nature works. We often think of life on Earth as made up of complex and tightly-knit ecosystems like rainforests, wetlands and coral reefs that are perfected and stable, with every species evolved to have a unique role. With that vision of nature, alien species are at best disruptive and at worse plain bad. But where did this idea come from? Darwin certainly never said it. He said natural selection allowed species to adapt and survive, but he said nothing about ecosystems evolving to some sort of perfect state. They were just a jumble of species making their way in the world.

And today, fewer and fewer ecologists believe nature is either stable or perfectible. Real nature, they say, is often random, temporary and constantly being remade by fire, flood and disease – with species coming and going, fitting in, adapting or losing out. Change is the norm, they say. In this vision of nature, alien species are just like any other. Whether brought by humans or in more traditional ways, they are not an intrinsic threat to ecosystems. They are part of nature doing its thing, constantly reordering itself, constantly submitting to random events. Aliens may or may not cause change, but if change is the norm then there is no harm in that. In any case, when invaded by foreign species, ecosystems don't collapse. Often they prosper better than before. The success of aliens becomes a sign of nature's dynamism, not its enfeeblement.

This new ecological thinking is critical for how we understand the meaning of conservation, and for what action to take in the name of protecting nature. If nature is perfected and vulnerable to outsiders, then conservationists have to man the barricades to keep out the interlopers and restore the balance of nature. That's what most conservationists think, and for a long time I shared that view. But if it is wrong, then keeping out aliens serves no obvious purpose. More than that, it may be counter-productive. Nature's desperados are proven colonists and exploiters of the ecological mess that humans leave behind them. So surely that makes them nature's best chance of healing the damage done by chainsaws and ploughs, by pollution and climate change. Far from being nature's destroyers, aliens may be its re-invigorators, its salvation. They may be a sign that nature is not done. That it can bounce back. If that thinking is right, then simple conservation is short-sighted and true environmentalists should be applauding the invaders.

I do not want to suggest that we should always welcome every alien species. We humans may sometimes want to protect the species we know and love – in the habitats that we are familiar with. There is nothing wrong with that. And where alien species cause us inconvenience – whether zebra mussels in American waterways, rats on oceanic islands or rabbits in Australia – we may want to try to halt their spread. Again, that is fair enough. We have a legitimate need to curb some of those excesses, and a legitimate desire to protect what we like best. But we should be clear that when we do this, it is for ourselves and not for nature, whose needs are usually rather different.

And while we seek to protect what we like in nature, we should remember something else. There is very little that is truly natural in nature any more. They are very few, if any, pristine ecosystems to be preserved. Thanks to the activities of humans over thousands of years, no forests are virgin. They are all regenerating from past human invasions. We live in a new geological era, the Anthropocene, in which nothing is undisturbed and most ecosystems are a hotch-potch of native and alien species, often getting on in unexpected and productive ways.

Over my years as a journalist, I have written plenty of articles about how much harm alien species appear to do – about killer algae, marauding water hyacinth and many more. There was truth in them all, but they missed the bigger picture. This book is my journey to discover what conservation should really be about in the 21st century. It should not be about trying to preserve nature in aspic, still less about trying to recreate the past. That is both impossible and an affront to nature, like trying to turn the world into a giant zoo. In the 21st century, rather than fighting a losing battle to protect what we imagine to be pristine nature, we should be encouraging nature's rebirth, often through the dynamism and invasive instincts of its alien species.

Nature is not there to do our bidding. While alien species may sometimes be a pain in the neck for human society, they are exactly the shot in the arm that real nature needs. Conservationists who want to cosset nature like a delicate flower, to protect it from the threat of alien species, are the ethnic cleansers of nature, neutralizing the forces of nature that they should be promoting. It is foolish to fear nature at its most dynamic – red in tooth and claw, rhizome and spore, root and branch. As true environmentalists, we should rejoice when species burst through the paving stones of our cities, or wash up on foreign shores. We should celebrate nature's powers of recovery. We should let it run wild. How else are species to thrive and respond to the disruption of our activities, including climate change, if not by invading new territories, by becoming aliens? True nature-lovers should see that.

Harvard biologist Edward O. Wilson, the guru of biodiversity and rainforests, said the 21st century will be 'the era of restoration in ecology'. I hope so. But we will not be going back to a supposedly pristine world. We cannot. We should be restoring nature's wildness, and not trying to turn one moment in the past into an ossified museum relic. The new wild will be different, but no less dramatic and wonderful than the old wild. Alien species, and the novel ecosystems they inhabit, will be at the heart of it. We should bring them on.

*

This book begins with a look at the reality behind some of the invasions by alien species that have made headlines. Part One starts with islands, where some of the most dramatic stories have emerged. Those stories tell of places where species introduced by humans have created healthy ecosystems where none existed, as well as of places where introduced species have ransacked colonies of sea birds. Usually, however, the stories are more nuanced. In case after case, I found that the supposedly malign invaders were simply taking advantage of ecosystems that had already been wrecked by humans. They were opportunists, but also nature's regenerators. They were often doing jobs that natives could not accomplish.

In Part Two, the story moves on to examine how our misplaced notions about aliens impact on the real world, and on how we do conservation. The results are often comical. Our efforts at ecological cleansing are rarely successful. They fail because conservationists have indulged ill-founded myths about aliens, pristine ecosystems and how nature works.

Having slain some myths, it is time to find some solutions.

Part Three attempts to reboot our ideas about nature. Most of the world is now composed of novel mixtures of native and alien species, happily getting along together, enriching our lives, maintaining ecosystems and recharging nature's batteries. This is the new wild. Nature is blossoming in the most unlikely places, such as logged-over forests and urban badlands. To make the most of that, we need to reboot conservation too. That means we need to lose our dread of the alien and the novel. It means conservationists must stop spending all their time backing loser species – the endangered and reclusive. They must start backing some winners. For winners are sorely needed if nature is to regroup and revive in the 21st century – if the new wild is to prosper.

PART ONE
ALIEN EMPIRES

All round the world, alien species are on the march, often with human help. But mostly they are moving into places we have messed up. They are often helping nature's recovery.

1
On Green Mountain

Standing on the summit of an extinct volcano on Ascension Island in the tropical South Atlantic, I was at one of the most remote spots on the planet. I watched the British military plane that had delivered me there, midway between Brazil and Africa, take off and head on south to the Falkland Islands. I felt rather alone. Down below was a harsh, black and treeless moonscape of volcanic clinker, baking in the sun. Beyond was ocean for a thousand kilometres in every direction. But in the cool mountain air, I was surrounded by lush greenery. As noon approached, a lone cloud formed over the summit and then suddenly descended, shrouding me and the mountain in mist.

The Ascension locals – a mix of British contract workers, American service personnel, and families from St Helena, another remote South Atlantic island – call this oasis Green Mountain. The island's British administrator has a bungalow up here, complete with a pair of old cannons pointing out across the ocean. But away from his lawns, where I later had afternoon tea, the mountain and its cloud forest felt primeval, a leftover perhaps from the days before sailors began visiting here five centuries ago.

My instincts couldn't have been more wrong, however. The greenery was relatively new. When Charles Darwin visited Ascension Island in July 1836, homeward bound after his long journey aboard HMS *Beagle*, he had complained about its 'naked hideousness'. The mountain where I now stood was 'entirely destitute of trees'. Another visitor of the day, William Henry Webster, had called the island 'an awful wilderness amid the solitude of the ocean'. Peering through the mist, my guide, the mountain's genial warden Stedson Stroud, explained: 'Nothing you see here is native. Except for a few ferns, everything has been introduced in the past 200 years.' The cloud forest of Green Mountain is an entirely man-made ecosystem, a pot-pourri of foreign species shipped in by the British Navy during the early and mid-19th century at the whim of Victorian botanists. Every passing ship had delivered more trees for the local garrison to plant.

On our way up the mountain, Stroud and I had walked through stands of Bermuda cedar, South African yews, Persian lilacs, guava fruit trees from Brazil, thickets of Chinese ginger, New Zealand flax, taro from Madeira, European blackberries, Japanese cherry trees and screw pines that grow taller here than at home on the islands of the Pacific. The summit was improbably covered in a dense stand of Asian bamboo, and rattled like a huge wind chime in the brisk trade winds that suddenly blew the mist away.

I was on Ascension because the very existence of this forest is creating controversy. It is more than a patch of trees, more than a botanical garden. It is possibly the most cosmopolitan tropical forest in the world, and it is said to be the only one that is entirely man-made. Moreover, researchers who have visited herald it as a fully functioning ecosystem, created from scratch in little more than a century from fragments assembled at random from around the world. The vegetation, insects and other species interact in numerous ways, providing vital services for each other. Forest ecosystems are not supposed to happen like that. Conventional ecology says their complex interactions emerge only as a result of long-term evolution of species. As Stroud put it, in a paper with David Catling of the University of Washington, the species on Green Mountain 'have bucked the standard theory that complexity emerges only through co-evolution'.[1]

Stroud had been tending the mountain for a decade, ever since he came here from St Helena to be the island's conservation officer. He admitted that, as a conservationist, he should probably be rooting out all those foreign trees, in order to allow the natives to regain their terrain. But if he did, there would be almost nothing left. And in any case, he said, he was presiding over something profoundly interesting. This confected cloud forest was prime evidence in a growing movement among ecologists to reconsider many of their nostrums about how ecosystems function. It suggested that species with no previous contact can get along with each other much more intimately than assumed. It suggested that perhaps many more forests and other complex ecosystems around the world are the result of chance meetings rather than complex co-evolution. If that is true, it may hold vital clues for regenerating nature in the 21st century.

Ascension Island, which is almost twice the size of Manhattan, erupted from the Atlantic floor a million years ago. It was not barren for long. Remote as it was, some life soon arrived. Millions of seabirds occupied the lowlands, their spattered excretions turning the piles of black clinker white. Green turtles swam thousands of kilometres to nest on its sandy beaches. And the island's mysterious land crabs arrived from who knows where – and who knows how – to make a life scuttling on the slopes of the volcano. But – apart from a few ferns and mosses on the mountain, and an endemic shrub called Ascension spurge along the shore – the black desert for a long time remained almost entirely devoid of vegetation.

The first known human visitors to Ascension Island were early Portuguese mariners who dropped anchor on Ascension Day 1503, on their way to the Indian Ocean. Hence the island's name. But the first permanent occupation was by the British Royal Navy. It set up a garrison in 1815, to patrol the surrounding oceans and prevent the rescue of the most famous international prisoner of the day, the former French emperor Napoleon Bonaparte, whom the British had captured and incarcerated on St Helena, its equivalent of Guantánamo Bay,

1,300 kilometres to the south. After Napoleon's death, the British used the island as a base for hunting down transatlantic slave ships, and to store fuel for warships heading to India, the 'jewel' in the British imperial crown.

Ascension Island has kept coming in handy. At the end of the 19th century, it became a hub for transatlantic telecommunications, with cables stretching to Europe, South Africa, Brazil and Argentina. These days, the island is peppered with aerials that track orbiting spacecraft, talk to nuclear submarines, and listen in secretly to cable and satellite-relayed communications. The electronic spies of Britain's GCHQ are the biggest employers. The island's band of temporary residents, who number about 800, say it has more antennae than people. It also has one of the longest airstrips in the world, built by the US Air Force during the Second World War to provide a secure stopping-off point for flights into Africa. When I arrived in early July 2013, I was amazed to see nine large US military aircraft sitting on the tarmac, all busy protecting President Obama during a visit to the continent.[2]

Inevitably, such human traffic has brought alien species, both accidental and deliberate. Recent arrivals on the lowlands include fast-spreading tobacco plants and Mexican thorn, or mesquite, which the BBC shipped here in the 1960s to brighten up gardens in a new settlement for operators of a transmission station serving Africa. Much was deliberately introduced. From the first, the naval garrison brought in species to make the remote outpost as self-sufficient as possible. Documents in Ascension's archive, in the toy-town capital of Georgetown, show how it first established a farm on one of the few patches of natural soil on the mountain. The farm grew introduced fruit trees like guava from Brazil, Cape gooseberries from South Africa, bananas from the Far East and lychees from China. Then there were vegetables such as cabbages, spinach and potatoes, as well as some grains, herds of South African pigs, and cattle and sheep from England.

The farm operated until the 1990s and is now overgrown. But it is the trees that fascinate in the 21st century. Britain planted on this volcanic hulk specimens from across its global empire. In the early days, the sailors grew New Zealand flax to make rope for ships, and straight-trunked Norfolk pine for masts. British colonial botanist Sir Joseph Hooker – a friend of Darwin and future head of the famous botanical gardens at Kew in London – visited in 1843. He came up with the idea of growing trees to gird the mountain and green the arid island. The archive preserves a letter that Hooker wrote in 1847 with a long list of recommended introductions to ensure that 'the fall of rain will be directly increased'. The new vegetation on the mountaintop would scavenge moisture from passing clouds, he promised. Further down the slopes, trees and bushes would encourage soil growth. Hooker's ambition was nothing less than remaking the volcanic island – or 'terra-forming' it, as Stroud and Catling put it.

In 1845, a naval transport ship from Argentina delivered the first batch of seedlings. In 1858, more than 200 species of plants arrived from the Cape Botanic Gardens in South Africa. In 1874, Kew sent 700 packets of seeds, including those of two types of plants that especially liked the place: bamboo and prickly pear. The sailors got to work, planting several thousand trees a year. The bare mountain was soon verdant – and renamed Green Mountain. An Admiralty report in 1865 praised the new cloud forest. The island 'now possessed thickets of upwards of 40 kinds of trees besides numerous shrubs', it said. 'Through the spreading of vegetation, the water supply is now excellent, and the garrison and ships visiting the island are supplied with an abundance of vegetables.'

Today, the island has around 300 introduced species of plants, says David Wilkinson, an eclectic botanist from Liverpool John Moores University who made a rare research visit to Green Mountain. Many are spreading. Above about 600 metres, Green

Mountain is now completely vegetated. On my walk back down the mountain, Stroud pointed out coffee bushes, vines, monkey puzzle trees, jacaranda, juniper, bananas, buddleia, palm trees, clerodendrum, the pretty pink flowers of the Madagascan periwinkle and the bell-shaped blooms of the American yellow trumpetbush. He confirmed that the vegetation captures more cloud moisture on the mountain, just as Hooker had hoped. This even though there has been a decline in rainfall in the lowlands around.[3]

The Victorian terra-formers did not just bring plants. The island's military inhabitants joined the 19th-century craze among expatriate Europeans to fill their new worlds with birds and animals from home. The regular ships to Ascension brought hedgehogs and rooks, ferrets and owls, bees to pollinate, and guinea fowl as quarry for hunting. Ascension never quite became the 'Little England' that they had hoped. Of the introduced birds, only the tropical canaries and mynas have stuck around. But the mammals did better. Rats and rabbits can still be found in numbers, along with feral sheep, cows and chickens let loose from the now-abandoned farm. And donkeys – descended from the beasts of burden that once carried water from mountain springs to the coastal garrison – still wander the landscape, eating mesquite fruit and getting hit by cars.

The introduced rats were trouble. They swiftly saw off a couple of local endemic birds, the Ascension crake and the Ascension rail, and possibly also a night heron.[4] For a long time, there were also feral cats. Originally brought in to control the rats, they tyrannized the seabird colonies, forcing most to nest instead on a tiny offshore mound known as Boatswain Bird Island. Ornithologists began a cat-eradication programme, and the last cat was hunted down in remote Cricket Valley in 2006. Since then, boobies have begun returning to the main island. When I visited, the island conservationists were crossing their fingers for the survival of the first nest established by a returning pair of endemic Ascension frigatebirds.[5] Meanwhile, 2013 saw a record nesting season for the green turtles – a great comeback after decades during which they were slaughtered for the pot. The last victim, they say, was fed to the Duke of Edinburgh when he visited back in 1957.[6]

The incomers have added hugely to the island's biodiversity and now make up at least 90 per cent of its species. The local people like them. When, in the 1980s, the island authorities issued twelve postage stamps bearing pictures of local insects, all twelve were aliens. The national flower, dubbed the Ascension lily, actually comes from South America. Nevertheless, some would like to encourage the resurgence of native species by launching an eradication programme against foreign plants. They blame the environmental anarchy perpetrated by Hooker and his successors for the loss of three endemic ferns, and perhaps some insects that may once have depended on them – though the lost species may still be hiding here somewhere amid the steep valleys around the mountain. As we stood on one of the mountain paths, gazing south over an abandoned NASA tracking station, Stroud pointed out below us the cliff face where, in 2009, he rediscovered a single specimen of a fern species believed lost, *Anogramma ascensionis*, or Ascension Island parsley fern. It is now being propagated in the labs at Hooker's old fiefdom in Kew, ready for reintroduction.

The British government's environment policy for the island is to carry out the 'control and eradication of invasive species' in order to 'ensure the protection and restoration of key habitats'. But this is dogma rather than considered policy based on research. A closer look would reveal that the 'control and eradication' of the aliens could now cause some natives, including endemic ferns, to go extinct. The 'key habitats' for some native species are provided by the alien trees on Green Mountain. *Xiphopteris ascensionis*, an indigenous fern that once clung to the bare mountainside, now lives only on the mossy branches of introduced plants. Bamboo is a favourite. 'The ferns like the shade,' says Stroud. 'If it weren't for the trees and other plants, I doubt if the ferns would be here any more.'

On our walk down the mountain, we saw native mosses clinging to an old stone wall, and alien insects and birds pollinating trees of a kind they had probably never encountered before arriving here. After the rains, we watched the distinctive yellow and pink land crabs – the island's largest native land animals – rush out of their burrows to feast on the fruits of alien trees like the guava. The only researcher to have studied these endemic land crabs in recent times, Richard Hartnoll of the University of Liverpool, says that the foreign vegetation 'increases the area of shade and shelter for crabs, and also provides a large resource of food' – perhaps replacing their former scavenging on seabird colonies. Removing the aliens would likely have some nasty consequences for the natives.

You might think that this ecological snugness among species thrown here from across the world would be of some interest to conservationists. Yet, until now, they have shunned it, says Stroud. Most of the handful of researchers who have made the long journey to Ascension – the only practical way is aboard a British military flight – have concentrated on the island's charismatic populations of seabirds and green turtles. A detailed study of the island's flora conducted in the 1990s by the University of Edinburgh catalogued only the natives and characterized the rest – the great majority – as simply a continuing threat to them.[7]

As a result of this wilful blindness, simple but important ecological questions go unanswered. 'As you walk through the forest, you see lots of leaves that have had chunks taken out of them by various insects', Wilkinson says. 'There are caterpillars and beetles around. But where did they come from? Are they endemic or alien? If alien, did they come with the plants on which they feed, or did they discover those plants on arrival?'

Wilkinson is among the scientists who propose that the

complex ad hoc interactions between native species and aliens from many lands on Ascension are good evidence for an ecological theory that contradicts mainstream ideas about co-evolution. Ecological fitting, a term coined by US ecologist Daniel Janzen, holds that ecosystems are typically much more random. 'The Green Mountain system is a spectacular example of ecological fitting', Wilkinson says. 'It is a man-made system that has produced a tropical forest without any co-evolution between its constituent species.' Thomas Jones of the US Department of Agriculture in Logan, Utah, agrees. On Ascension, he says, plants gathered from across the world 'self-organized by the mechanism of ecological fitting'.

This debate matters. It raises in practical form many of the questions I will explore in this book. What are ecosystems and how do they form and function? Are alien species 'bad' and natives 'good', or is this distinction false and scientifically unjustified? Could nature be far more resilient, and far less fragile, than we imagine? Have we messed with the environment so much now that places like Green Mountain, with their bizarre mixtures of natives and aliens, offer the best chance for nature's survival in the 21st century and beyond? Is the success of aliens the most vivid expression of Darwin's 'survival of the fittest'? Should we – for the good of the natural world – learn to love the aliens?

*

Some regard such talk about alien species as nonsense. And it is true that not all alien species fit in like respectable ecological citizens. Some create bloody mayhem, especially within the simple ecosystems that occupy many remote islands. Gough Island is another volcanic hulk far out in the South Atlantic. Like Ascension almost 4,000 kilometres to the north, Gough Island was first discovered by Portuguese navigators but claimed only much later by the British. They raised the Union Jack there in 1938 and named it after Charles Gough, the first British sea captain to spot the island, 200 years before. Gough has no natural harbour. Climbing ashore is difficult. As a result, sailors have only ever set foot here about 250 times. That includes the seasonal change of its only current inhabitants, the South Africans who man the weather station. Yet 71 of the 99 species recorded here were introduced during those 250 visits. One of them is a big problem.

For as long as anyone knows, Gough Island has been dominated by millions of nesting sea birds. When Britain nominated the island as a World Heritage Site back in 1995, the Royal Society for the Protection of Birds (RSPB) called it probably the world's most important seabird colony. Its 10 million avian inhabitants included 90 per cent of the endangered Tristan albatrosses (*Diomedea dabbenena*) and the entire world populations of both the ground-nesting Gough bunting and the flightless Gough moorhen. The island is now also the only known breeding ground for endangered Atlantic petrels (*Pterodroma incerta*). (The other colony, on Tristan da Cunha, got eaten out by invading rats.)

There are no trees on Gough Island, and vegetation is mostly lichens and ferns. So to escape from the wind and cold, the birds live in burrows. But this subterranean seabird megacity is under threat. For there are new masters on Gough Island – the descendants of English house mice (*Mus musculus*) that leapt ashore from Victorian whaling ships sometime in the 19th century. These are not normal mice, however. Not now. Over the decades of their windswept exile, they have doubled in size and turned carnivorous. These mutant mice now typically grow up to 25 centimetres long and eat up to a fifth of their body weight every day. With an estimated 2 million mice on just 6,500 hectares, that means they devour three tonnes of bird flesh every day from every hectare of the island.[8]

Cape Town ornithologists first detailed the carnage during a visit in 2001. They calculate that only a quarter of the 1.6 million petrel chicks that hatch each year on Gough make it to adulthood. Virtually all the rest are eaten by mice as they sit immobile in their underground nests. Even albatross chicks, which may be the size of geese, succumb to the night-time attacks by these marauding monsters, says an incredulous Ross Wanless, who has taken video of the gruesome scenes.[9] 'It's like a tabby cat attacking a hippopotamus', says Geoff Hilton of the RSPB, which wants to exterminate the mice.

How do you get rid of 2 million mice on a remote island? A feasibility study recommended hiring a helicopter to bomb the place with some 100 tonnes of brodifacoum, an anticoagulant poison widely used against rats, at a likely cost of £1.5 million. Even then, since many of the mice live in caves and volcanic lava tubes, complete eradication could not be guaranteed. And there was a risk the poison could instead end up killing off the Gough moorhen. Perhaps not surprisingly, the enterprise was on hold in early 2014.[10] The supermice remained in charge, and may remain so until they run out of food.

Despite their atypical nature, simple island ecosystems such as Gough have become the test cases for what we think about aliens – perhaps because, surrounded by ocean, they provide understandable terrain for these battles to play out. The most discussed of all sites of invasions is Hawaii, where both those vehemently opposed to alien species and those who see their virtues are willing to make their cases. On one side is Daniel Simberloff of the University of Tennessee. A guru among invasion biologists, he is now in his 70s. He studied half a century ago under Harvard's legendary forest ecologist and pioneer of island biogeography Edward O. Wilson. He says Hawaii is the site of an 'invasional meltdown'.[11] On the other side is Joseph Mascaro, a post-doc at Stanford University, 40 years younger than Simberloff, whose work in Hawaii was publicized in the groundbreaking book *Rambunctious Garden* by journalist Emma Marris. Mascaro says Hawaii is simply exhibiting some textbook examples of what he and other ecologists are now calling 'novel ecosystems'.[12]

The Hawaiian Islands are the most northerly of the Polynesian islands in the central Pacific. They are an archipelago of volcanic islands. One volcano, Mauna Loa, still spews lava today. Others are long extinct and have been above the waves for 10 million years or more. Species reached the islands on the winds and waves, in the gullets of migrating birds and clinging to the trunks of floating trees. Coconuts came bobbing by. They all made themselves at home as best they could. Lichens turned the lava to soil; seeds took root; birds spread the seeds of plants. Evolution hit overdrive. Genetic mutations created new forms through the constant rolling of DNA dice; and the successful flourished. More than a hundred bird species evolved from maybe twenty arrivals. Finches, probably from Asia, evolved into 56 different species of endemic Hawaiian honeycreepers, almost half of which are still flying around, pollinating both native and alien vegetation.

Many other unexpected things happened. As environmentalist Louise Young noted in her book on islands: 'Wingless and blind insects evolved … land snails became tree snails … unusual varieties of blooming trees [emerged] like lobelias, hibiscus and tree violets.'[13] But this largely random collection of species also contained some big gaps. 'Hawaii lacked reptiles, amphibians, flightless mammals and ants', says Mascaro.[14] Pollinating insects never got going there, leaving the role to the remarkable collection of birds.

Human interference on Hawaii probably began when Polynesian canoes rode ocean currents from the Society Islands onto the beaches of Hawaii some 1,500 years ago. The newcomers cut down lowland forests for farmland, while leaving intact the upland forests, which they regarded as sacred. They exterminated some meaty flightless birds. But they brought the first freshwater fish, as well as rats and pigs and the candlenut tree that they used to make canoes and provide nuts for lamps. The newcomer humans prospered, as did many of the species they had brought. When the English adventurer Captain James Cook first visited in 1778, there were several hundred thousand Polynesians there.[15]

Europeans brought many more alien species, like cats and mongooses to hunt down rats; pigs and giant African land snails for food; deer for hunting; guava and other useful trees. Out went a number of the islands' unusually large range of bird species, including the Hawaiian rail, last seen in 1884. Nonetheless, each new invasion added biodiversity, since many more species arrived than were made extinct. Hawaii contains some 1,500 species of flowering plants found nowhere else, but also more than a thousand new plants have arrived. There are only 71 known extinctions. The estimated 2,500 introduced insect species have added 50 per cent to the native component. The avian tally is 66 bird species lost and 53 gained, though most of the losses were hunted to extinction by Polynesians before Europeans arrived. Some of the introductions were made to combat pests, like the worm-eating myna bird from Asia. Others, such as song birds and the American northern mockingbird, a notorious mimic, probably won passage for the sheer joy of their company.[16]

Whether adornments or not, most newcomers also provide ecological services. Alien birds such as the Japanese white-eye and the red-billed leiothrix, introduced from India a century ago, are today the mainstays for dispersing the seeds of native shrubs. 'Introduced mangroves are straining sediment and building habitat that native fish utilize', says Mascaro. Most ecologists do not spot these important services, he says, because they dismiss ecosystems containing alien species as damaged and degraded. Thus, they are beneath the attention of those who are searching for the truth about nature. This, he says, is a big mistake, because much of the truth about nature is bound up in these disrupted ecosystems.[17] Getting rid of the alien birds would leave the understorey of many native Hawaiian rainforests bare, says Jeffrey Foster, now of the University of New Hampshire.[18]

Most controversial is the case of *Morella faya* (formerly *Myrica faya*), a myrtle tree sometimes called the 'fire tree'. It was brought to Hawaii by Portuguese migrants a century ago from Atlantic islands like the Azores. It enjoys the sometimes harsh volcanic conditions and is the first plant to colonize newly created lava flows in the Hawaii Volcanoes National Park. It 'fixes' nitrogen from the air and uses it to turn the lava into soil, which no native plant can do. Many ecologists nonetheless think this skill is a bad thing. Princeton's Andrew Dobson concluded in 1998 that by colonizing the lava flows, the myrtle 'shuts out native species and leads to their subsequent extinction'.[19] Ecologist Peter Vitousek of Stanford University took the same view, though he agreed that their nitrogen-fixing skills might eventually turn the lava into soils fit for native plants.[20]

I thought that, more than a decade on from these pronouncements, it would be interesting to know what happened. Did the fire tree pave the way for other species or crowd them out? But there is a dearth of published data. Nobody, it seems, has bothered to research the question. When I checked the International Union for Conservation of Nature's Global Invasive Species database, it had no references to research into the impacts of *Morella faya* on Hawaii since 1991.[21] Mascaro told me that from what he saw, the myrtle appeared to be receding from lava flows as native species colonize the soils it has created. He said the case was a perfect example of how conservationists can, through their love of native species and ecosystems, simply fail to recognize the services that aliens often provide to the wider environment. Novel ecosystems work.

*

Odd things happen on islands because their simple ecosystems have weak spots and tipping points beyond which dramatic changes can take place. Sometimes the arrival of outsiders triggers these tipping points. In the Indian Ocean, on Australia's Christmas Island, the dynamic new ingredient was yellow crazy ants (*Anoplolepis gracilipes*) from West Africa. The ants have spread widely across the tropics, hitching lifts with passing trading ships, and are reckoned to have been on Christmas Island for the best part of a century. But they suddenly went on the rampage in the 1990s, creating 'super-colonies' of a kind never seen before. Nobody is sure why this happened, but they may have benefited from the spread of sap-sucking insects, since the ants eat the honeydew secreted by the insects. The ants and sap-suckers certainly seem to have formed a strong reinforcing relationship that ecologists call 'mutualist', even though one is alien and the other native.

The ant super-colonies extended for hundreds of hectares and contained billions of insects. As they grew, things got out of hand. The ants came into ever greater contact with the island's previously dominant species, red land crabs (*Gecarcoidea natalis*), during their annual migration to the coast to breed. The ants adopted their normal defence mechanism, spraying formic acid around. The acid blinds and eventually kills the crabs. Since 1995, some 20 million crabs have been killed by the ants, equivalent to around a quarter of their total population. Though not much more than a tenth of the island has been invaded by ants, their impact has cascaded through the island's ecosystems. Where the crabs have been eliminated, many more forest seedlings germinate, and parts of the forest have seen a massive increase in foliage in the understorey.[22]

Meanwhile, with fewer crabs to eat them, another long-established invader, the rat-sized African land snail, is now also on the march across the island.[23] And the 'crazy' ants – so-called because of their long-legged physique and frenetic behaviour – have won an instant place on the global list of the hundred most invasive species, drawn up by the IUCN.[24] Actually, it is far from clear that they should be blamed. What caused the super-colonies to explode so suddenly after the ants had been on the island for so long? Nobody knows. But that may be the heart of it.

Ants can travel the world easily, and often thrive quite well on arrival – in part no doubt because they arrive in crowds. One recent study suggested that at least 600 different ant species may have made themselves at home in alien environments around the world. Notorious arrivals include South American raspberry crazy ants, now swarming inside electrical equipment in the US, and Argentine ants (*Linepithema humile*), which have been building super-colonies in Europe.[25] But while invader ant populations sometimes boom, they also usually bust. The Argentine ant arrived in New Zealand in 1990 and marched across both the north and south islands in short order, killing chickens, invading orchards and pushing aside native ant species. 'They can even squirm under the edge of screw top jars and follow the grooves until they reach the contents', a government public warning advised. Officials set aside around $60 million a year to confront the ant army. Then in 2011, most of the known super-colonies collapsed. Why? Nobody is sure. But the panic is over, and native ants are recovering.[26]

*

Many remote islands have nurtured unique collections of species. The Seychelles in the Indian Ocean, for instance, have been isolated for more than 60 million years, gathering passing wildlife and turning the itinerant species into their very own. Aldabra, one of the Seychelles' outer islands, is home to flightless birds like the white-throated rail, and some 100,000 giant tortoises, two-thirds of the world's total. Aldabra remains remarkably unaltered, but the native species on many such islands have suffered. They have nowhere else to go, and the simple island ecosystems are wide open to outsiders with skills unknown to the naive locals. More than 90 per cent of the hundred-plus birds known to have become extinct over the past 400 years were endemic to islands, according to the World Conservation Monitoring Centre in Cambridge. Though many were hunted to extinction by humans, introduced predators have certainly taken their fill. Notoriously, ground-nesting birds with no experience of mammalian predators are suddenly confronted with mega-mice, cats or, most frequently, rats.

The black rat (*Rattus rattus*) has been causing trouble ever since it left India for Europe some 5,000 years ago. If nothing else, it brought Europe the fleas that carried the Black Death. The Polynesian rat (*Rattus exulans*) has been similarly troublesome across the Pacific. It is now being blamed for the mysterious crash of the once grand civilization on Easter Island in the remote south-east Pacific, whose collection of giant stone statues of humans has long mystified explorers. The theory used to be that the Polynesian Rapa Nui people who made the statues deforested their island to death. With the trees gone, the island ecosystem collapsed, and the islanders could no longer make the boats they needed to go fishing. The once-proud island people were reduced to a small emaciated rabble. It was ecocide, a warning to the world about the folly of mankind's environmental sins. Jared Diamond asked in his book *Collapse*: 'Are we about to follow their lead?'[27]

But now it seems that when the Rapa Nui arrived on the island a thousand years ago, they also brought seed-eating Polynesian rats. Rather than humans destroying the forests with axes and fire, it was the rats that delivered the *coup de grâce*, says Terry Hunt, an anthropologist at the University of Oregon. The rats ate the seeds of the island's palm trees, so the trees did not reproduce. Virtually every palm seed shell dug up on the island has been gnawed by a rat, Hunt says. 'It was rats, more than humans, that led to deforestation.'[28]

Henderson Island is a British-run World Heritage Site in the

Pacific that has been uninhabited since the Polynesians left in the 1600s. Unfortunately, they left behind Polynesian rats, which have overrun the place and are currently eating their way annually through some 25,000 seabird chicks, including endemic petrels. A million-pound effort by ornithologists to poison the island's rat population in 2011 failed. Zoological explorer Mike Fay, whose exploits in Africa we will hear more about in Chapter 8, spotted one while tramping across the island the following year.[29] A reconnaissance group sent out by the RSPB in 2013 found that a small population was still there, probably happily breeding. RSPB's Jonathan Hall told me there would be 'a further eradication attempt'.

More rarely, snakes turn small islands into horror movies. Until 1950, the only snake on Guam in the Western Pacific was the Brahminy blind snake, a small slow-moving worm-like reptile that could not see where it was going and mostly lived in termite mounds. Then the American military showed up. Somewhere aboard the ships and planes moving equipment from wartime bases on the Admiralty Islands of Papua New Guinea to the new US Pacific headquarters on Guam, came a contingent of brown tree snakes (*Boiga irregularis*).[30]

On their home islands, brown tree snakes were just one element of a diverse reptilian menagerie. But on Guam they had an electrifying effect from the moment they slithered off into the bushes. They proliferated fast, colonizing half the island by the mid-1960s and most of the rest by a decade later. In places, there were soon 100 snakes per hectare. Fanged and venomous, with large head and bulging eyes, they were very different from the blind and harmless native. They travelled everywhere, climbing trees and invading buildings during their nocturnal hunting expeditions. They ate almost anything, from dog food to lizards and birds to animal carcases. Ten of the island's native forest bird species, along with several species of native bats and lizards, went locally extinct after the arrival of the brown tree snake. Brought in to work out why, avian pathologist Julie Savidge concluded that the snakes were to blame. Gordon Rodda, who works on invasive snakes for the US Geological Survey, says there could at one time have been as many as four snakes for every bird.[31]

The snakes' notoriety grew. They got a reputation for entering houses and taking a nip at babies in their cots. There were more than a hundred hospitalized cases.[32] The invaders climbed wooden poles carrying power lines, presumably mistaking them for trees, and slithered along the wires looking for snoozing birds. Growing up to three metres long, they were heavy enough to drag down the lines, causing hundreds of power cuts.

There are many theories as to why this previously little-noticed snake caused such a stir on Guam. It certainly stumbled on a place where snakes were a novelty. Local victims had no idea how to defend against the newcomers. But equally, it may have found an island ecosystem in crisis and wide open to takeover. Guam had only fragments of its old forests. Most had long-since been cleared for coconut plantations. Much of the island had been bombed by the US and Japanese as they fought over it during the Second World War. It had then been washed with DDT to rid it of malaria. By the time the carpetbagging snake showed up, the remains of the native forests were rapidly being cleared for the runways, houses, golf courses and other infrastructure needed to satisfy tens of thousands of US troops on an island less than a sixth the size of Long Island.

Such changes – by disrupting existing ecosystems, removing other species and creating numerous ways for the incomers to prosper – must surely have contributed to both the wildlife decline and the conditions that allowed the remarkable spread of this hungry and crafty snake. The snake cannot be blamed for the loss of several species of birds that had all but gone before it showed up. The nearby US-run island of Rota, which was at the time free of snakes, has shown a similar decline in native forest birds, which ornithologists attributed to habitat loss.[33] Without the snake, there would undoubtedly be many more birds on Guam today; but without the massive disruption caused by the arrival of the US military in particular, the snake is unlikely to have prospered so well.

The heyday of the brown tree snake on Guam may be past, however. Its numbers have fallen to roughly half the 4 million estimated during its marauding 1980s peak. Perhaps it has simply been too successful in eating the available food. But the US military is taking no chances. In spring 2013, it dropped thousands of dead mice into the forests of Guam. Each mouse carcase was stuffed with tablets containing paracetamol and kitted out with a tiny parachute made of green tissue paper. The mice floated into the branches, where the plan was for them to be gobbled up by the snakes, for whom the painkiller is lethal.[34] Will the programme succeed in wiping out the snake? We shall see.

Stories from places like Guam and Gough Island have created a widespread assumption that oceanic islands are sitting ducks for disaster once alien species arrive. Many also conclude that islands reveal the worldwide effect of aliens in a microcosm. Neither is true. Isolated islands have very singular ecosystems and there is no reason to think the rest of the world will be like them. And when Christoph Kueffer, an ecologist at the Swiss Federal Institute of Technology in Zurich, looked at over 250 'invader' plants on 30 island groups from Ascension to the Virgin Islands, he found that only a handful of the invaders – out of tens of thousands of known introductions – were of much consequence for local species. Most alien species add to their local diversity and enrich their species-poor ecosystems. There are many more islands like Ascension and Hawaii than like Gough or Guam.[35]

2
New Worlds

Australia is the world's biggest island. Virtually cut off from the rest of the terrestrial world for millions of years, it was a place of evolutionary throwbacks like marsupials and the egg-laying duck-billed platypus, a creature so strange to outsiders that the first specimens taken back to Europe were dismissed as a hoax.[1] Until the first Australians arrived from South Asia more than 40,000 years ago, it had giant emus, marsupial lions that could stand on their back legs, a spiny anteater the size of a pig, a horned tortoise, its own giant lizard and 2-tonne wombats. The reason for the demise of these bizarre mega-creatures is disputed, though the early human migrants are generally blamed. Some of the animals may have been hunted down, while others probably succumbed to the massive fires that the new humans are believed to have set as they pursued and corralled their prey and burned grassland to encourage regrowth.

This is not the place for a detailed history of Australia's ecology, but three introduced species embody many of the profound changes that have accompanied recent human influences on the Australian landscape and ecology: the dingo, the sheep and the rabbit. Their stories tell an important truth about what happens when we take species round the world. We expect those species to transform our new worlds into replicas of the worlds we left behind. And when they are not so obliging, we blame them. When ecosystems change dramatically, or simply fail to deliver our needs, we look for scapegoats, and usually blame our alien companions. The truth is always more complex. In the melting-pot landscapes that normally follow human mass migrations, almost everything changes. There are usually more species than there were before. New arrivals far outnumber any resulting extinctions. Local biodiversity increases. We may mourn what is lost. But there is no going back.

The dingo (*Canis lupus dingo*), Australia's wild dog and the country's largest land predator, was originally not so wild and not so Australian. But it is the oldest animal that we know with any certainty to have been introduced by humans. It probably arrived aboard seafarers' ships from Asia some 5,000 years ago, initially as a domesticated companion. Aborigines adopted it to help with hunting. But it also took off into the wild, where it prospered.[2] It was soon top dog, seeing off rival native carnivores like the thylacine, often called the Tasmanian tiger. The thylacine was finally confined to exile on the island of Tasmania, where humans eventually hunted it close to extinction, before the last dog died in a Hobart zoo in 1936.

Nobody seems quite sure whether to regard the dingo as alien or an honorary native. It touches a raw nerve in Australia. Some love the animal and believe it can do no harm. That resulted in a famous miscarriage of justice. Lindy Chamberlain was convicted of killing her child at a remote outback camping site in 1980 because the court believed expert assertions that her cries that 'a dingo's got my baby' were simply impossible. It took years for her conviction to be overturned. Others simply do not like the animal. 'Dingo' is Australian slang for a coward.

Some reckon that after 5,000 years it should be regarded as a native. The New South Wales state legislation recognizes it as a native dog.[3] Conservationists have taken a similar view in working to prevent the genetically 'pure' Australian dingo from interbreeding with domestic dogs. There may be some complicated cultural associations going on here. 'The dingo narratives affirm the notion that colonists can be indigenized', says Adrian Franklin, a sociologist at the University of Tasmania, in his book

Animal Nation. In other words, it makes Europeans feel less like colonists and more like genuine Australians.[4]

When European colonists established permanent settlements in Australia in the late 18th century, they tried to make their new land productive, and introducing species was the method they chose. Most introductions failed. In the 1860s, Peruvian adventurer Charles Ledger smuggled more than 300 alpacas over the Andes to Chile and then shipped them across the Pacific to Sydney, where the New South Wales government paid £45 a head for them. They all died soon after. Imports of agoutis and Angora goats also failed. In neighbouring Victoria, the list of failed introductions included llamas, eland and birds such as linnets, yellowhammers, robins, grouse, quail and ostriches.

But there were successes. Camels brought in the 19th century for construction work were set free once trucks took over. There are today more feral camels in Australia than there are camels of any stripe in the deserts of Arabia. The English shipped foxes to Australia so they could hunt them. But instead the foxes went hunting themselves, taking koalas, marsupial rodents known as bilbies, birds and their eggs, mice, frogs, fish, lizards, bats and much else. The landscape came to resemble parts of Europe. English novelist Anthony Trollope was a fan. In 1873, after visiting his son, a sheep farmer in New South Wales, he noted how 'In some districts … the English rabbit is already an almost ineradicable pest; in others is the sparrow. The forests are becoming full of the European bee. Wild horses roam in mobs of thousands …'[5]

By 1890, Australia also had 100 million sheep (*Ovis aries*). Their grazing changed the pasturelands. Sheep grazed and trampled the grasses that held the soil together. Australia had previously been grazed by animals with padded feet that tripped lightly across the grasslands. The newcomers had hooves – 'hobnails rather than rubber soles', as author William Lines put it in *Taming the Great South Land*. As a result, the clay cracked in the sun and the hillsides collapsed. Meanwhile, farmers cleared the land of its dry brush and forest to make more room for the sheep. They planted European grasses, and even where they did not, their sheep distributed the seeds of alien plants carried in their wool all the way from Europe. Ryegrass, cocksfoot, white clover, furze and thistle all spread in that way, says Lines, replacing Mitchell grass and other natives. Sheep, and the changes that their owners made to accommodate them, changed Australia. Yet virtually nobody blames the sheep for the state of Australia today. They blame the rabbit, which farmers hated because they ate the grass planted in arid areas for sheep.

When an English settler named Thomas Austin took 24 European rabbits (*Oryctolagus cuniculus*) to his ranch near Geelong in 1859, he wanted nothing more than to feel at home. The shipment also included hares, partridges, blackbirds and thrushes. But the rabbits liked their new home best. They bred like, well, rabbits. As their numbers grew, Austin invited shooting parties to hunt them. But the animals outran the guns and escaped. Soon there were millions of them. Rabbits were undeniably aggressive adventurers, racing north through Queensland by the 1880s, defying bounty hunters and the best efforts of the Intercolonial Rabbit Commission. They easily traversed the Nullarbor desert, and bust through a 2,000-kilometre fence erected to keep them out of Western Australia.

Even the great drought in eastern Australia in the 1890s only slowed their progress. With the pastures gone, they swarmed in the cool of the night, reputedly piling up at fences until there were enough bodies for the next wave to climb over and keep going. By 1950, despite a thriving trade in tinned rabbit meat and rabbit skins, there were half a billion of them; then the government introduced a new alien, a Brazilian rabbit disease called *myxomatosis*, which killed all but a few thousand. The survivors, however, became the basis for a revival.

Rabbits were blamed for turning rangelands to deserts. And

they took the rap for a multitude of other crimes like taking over the holes of kangaroo rats, leaving them defenceless against foxes. But while the tale of the rabbit is almost a defining myth of the new nation, it is largely scapegoating. The real damage had already been done by the farmers and their sheep. 'In Australia', says Mark Davis of Macalester College, 'declines in the native species typically began decades before introductions of species ... that have been reported to have contributed to the extinctions.'[6] By trying to turn Australian grasslands into European sheep farms, the colonists had 'built a rabbit paradise', says environmental historian John McNeill. In the end, the rabbits couldn't fail. They simply occupied the ecological space created for them.[7]

*

Humanity's involvement in moving species around the world is as long as our own migrations. Tribes of *Homo sapiens* left Africa for the Middle East 125,000 years ago, before the last ice age. They went to South Asia 50,000 years ago; to Europe and Australasia 40,000 years ago; and to the Americas perhaps 12,000 years ago. We took species for company and food, to clothe our bodies and ornament our gardens, for construction and manufacture, to hunt and fight pests, for zoos and as pets.

Many species have been eager to share our journey. We took with us, either deliberately or accidentally, many of the same companions we have today: dogs and cats; rats and mice; earthworms and bees; sheep and goats; starlings and pigeons; wheat and barley; bananas and beans. They have been central to our conquest of the planet. Before we invented agriculture, only a few million humans lived on Earth. But domesticating and transporting species of plants and animals allowed us to feed many more. Today 7 billion of us are fed each day, mostly from plants grown very far from where they originated. The wild varieties of wheat grew in the Middle East, maize in Central America, potatoes in the Peruvian Andes, soy in China, and rice in South-east

Asia. Polynesians sailed the Pacific with pigs, sweet potato, breadfruit and taro – and took chickens to Chile about a hundred years before Columbus began the European species translocations.

Those transatlantic migrations, dubbed the 'Columbian exchange' by environmental historian Alfred Crosby, have always been two-way traffic. Wheat went to the Americas and potatoes and chocolate came east to Europe. African slaves took rice with them to the Americas, while the British extracted rubber, cinchona (to fight malaria) and peanuts from the Amazon jungles. There were other global journeys. Acacia trees first arrived in Australia with Pacific traders, having originated in the Americas. They then spread on colonial whims across the world, including to South Africa, where some of the largest plantations were established.[8]

We can sometimes forget how much of our everyday landscape is foreign. Travel around the Mediterranean and you may imagine that the landscape is composed mostly of native crops. But the citrus trees were brought by Arab traders from the Far East, the cypresses are from Iran, the tomatoes and tobacco, cactus and chilli peppers and much else from the Americas. Cattle crossed into Spain from Africa some 4,000 years ago.[9] Even the most remote of African farmers are often growing crops from far away. Arab seafarers took bananas from New Guinea to Africa around 2,000 years ago, probably along with yams. Avocados, plantains and pineapples crossed the Atlantic from the Amazon and penetrated up its long rivers deep into Africa. Mangoes came from South Asia, cassava from Central America and goats from the Middle East. Malaysians first brought rice to Madagascar a thousand years ago.

We live in what Christian Kull of Monash University calls 'melting pot landscapes'.[10] Local ecosystems of course remain distinct and limited by geology and climate. But species come and go so much, as a result of both human and natural forces, that conventional hard distinctions about what belongs where have long since become all but meaningless.

So much has been moved so many times in much of Asia, for instance, that nobody would even try to make the distinction. Such is the confusion that even native Asian weeds are sometimes thought of as foreign. After American forces sprayed millions of tonnes of Agent Orange to defoliate rainforest during the Vietnam War, a tough local grass called cogon moved into the clearings. It provided ground cover until the native trees got going again. But locals associated it with the bombardment and called it the 'American weed'. It was never an American weed – though it may soon become one. It has since invaded the US in the packaging of houseplants from the region.[11]

Or take the ecology of the island of Borneo, shared today between Indonesia and Malaysia. Until recently, it was regarded as one of the most wild and untamed regions of Asia, an almost mythological jungle-infested place, visited by daring adventurers and the supposed home of circus freaks like the 'wild men of Borneo'. Yet all is not what it seems. Many of its endangered orang-utans, though feted as symbols of the untouched rainforest, are probably the descendants of animals kept captive for their meat centuries ago.[12] Whereas the pigs and cattle farmed there turn out to be mostly the descendants of wild natives. The omnipresent water buffalo may look local, but was actually introduced from the Asian mainland after being domesticated some 5,000 years ago. Terrifying-sounding fish like the snakeheads are domesticated and have been farmed in irrigation canals for at least two millennia. And many birds fluttering in the forest canopy have turned wild after being transported as caged birds and then escaping. The Java sparrow is one example.

Farmers in Borneo, as in most places, grow mostly foreign crops. Rice came from southern China; coconuts from the Western Pacific islands; black pepper, from the Western Ghats in India, was brought to Indonesia by Hindu colonists 2,000 years ago. The swamp sago, a long-standing staple food here, has made many human-sponsored journeys through the region. More recently, peanuts came to Borneo from Bolivia, and coffee and cassava from Africa, while Europeans brought plantation crops like rubber, acacia and eucalyptus. A local culinary speciality is the American bullfrog (*Lithobates catesbeianus*), caught in the canals of the city of Kuching.

Most such introductions worldwide were intentional. But many were accidental. Small creatures and plant seeds found it easy to slip ashore unnoticed amid ships' cargo. Vessels arriving in Virginia in the early 17th century gave a ride to rats, which thanked their new hosts for their hospitality by eating their way through grain stores and leaving the settlers starving.[13] Ships reaching New England around the same time brought the first earthworms to live in soils there since the last glaciation. The pioneers probably wriggled free from soil loaded aboard in Cornwall in south-west England as ballast for the journey. They soon began eating leaf litter that had previously built up on the floor of woodlands, permanently changing forest ecosystems.

Nobody takes much notice of earthworms' movements, not even biologists. We know little about their global journeys, but they have undoubtedly been extensive. More than a hundred species are widespread around the world, probably mainly thanks to humans. Paul Hendrix of the University of Georgia, who hosts a group called the International Symposium on Earthworm Ecology, says: 'introduced exotic earthworms now occur in every region in all but the driest or coldest habitat types on Earth.'[14] Their effects can be regarded as good and bad, aerating soils but also changing soil chemistry and ecology. They travel accidentally in soil, or more deliberately through the surprisingly large global trade in fishing bait. Canadian exports of fish bait alone are worth $20 million a year: that is a lot of worms. More bait comes from Europe, Japan and Vietnam.

Among the most unruly of the accidental migrants have been diseases and pests. European travellers returning from Asia in the 14th century brought home black rats that carried the bubonic plague, ultimately killing some 75 million people. The first colonists of the Americas wiped out most of the inhabitants with smallpox, measles and other alien diseases that they carried. It was 'the worst human holocaust the world has ever witnessed', University of Hawaii historian David Stannard argues.[15] But they didn't return empty-handed. They brought home syphilis.

People-trafficking to provide labour for the British Empire carried hookworm from India to the Caribbean and cholera to the Far East. A North American fungus, *Phytophtora infestans*, infected the Irish potato crop in the 1840s and, with the help of British administrative heartlessness, killed a million Irish in the subsequent famine. A decade later, the American vine aphid, *Phylloxera*, reached the vineyards of France, almost wrecking the wine industry, until someone thought of grafting European vines onto American rootstock resistant to the pest. And, as the century closed, Italian troops took to East Africa cattle that carried rinderpest, an Asian virus similar to canine distemper that, as we shall see in Chapter 8, changed the continent's ecology for ever.

Some journeys by pests defy prediction. European airport arrivals halls often carry warnings about bringing in the Colorado beetle. In the 17th century, *Leptinotarsa decemlineata* was living quietly in central Mexico, eating the seeds of the Mexican burweed (*Solanum rostratum*). Then Spanish conquistadors brought in cattle, which they drove north to sell in Texas. The animals carried burweed seeds caught in their coats, and the beetles came too. Liking the lie of the land, burweed and the attendant beetles moved north again into the Great Plains.

In Colorado, the beetles encountered fields of potatoes, which also had made a long journey to get there. The vegetable had been taken from the Andes to Europe, and subsequently introduced to the US in the early 18th century. In Colorado, the beetle swiftly developed such a taste for the potato that it switched hosts. Finding potato fields everywhere, it moved east across the US, before making it to Europe via US military bases set up near Bordeaux after the First World War.

In the early 1950s, Soviet-occupied potato-eating East Germany became heavily infested with the Colorado beetle, leading to charges – probably erroneous – that the 'Yankee beetle' was being dropped from US planes as a biological weapon. The Cold War briefly threatened to become a hot potato war. Tempers cooled, but the beetle became established in the Soviet Union by 1955 and is to this day heavily quarantined to keep it out of Western European countries.[16]

*

For Europeans, there was both commercial zeal and scientific curiosity in the desire to move stuff around the world. In the 1680s, Hans Sloane, the personal physician of Lord Albemarle, the British governor of Jamaica, returned home with the embalmed remains of his deceased employer. Also in his packing cases was a huge collection of forest goods from the Caribbean, including 800 plants and a two-metre yellow snake, subsequently shot by Lady Albemarle after it escaped from a pot. Sloane took more care of his coca plants, however, turning them into a new career. Having seen Jamaicans prepare a drink made from coca, honey and pepper, he came up with the idea of mixing the ingredients with milk, which he sold in London as 'Sir Hans Sloane's Milk Chocolate'. Thus began the British fixation with chocolate.[17]

Sloane, who later gave his name to Sloane Square in London, continued collecting plants, from opium poppies and cannabis to Chinese rhubarb, as well as a variety of insects and thousands of objects from around the world. His treasure trove, including a large herbarium, became the basis for the British Museum collection. Great private collectors of plants and animals of the Victorian era included Lionel Walter Rothschild, of the banking family, who created a huge museum and animal park in Tring, north of London. From there, as we shall see later, a number of celebrated animal escapees made their first appearances in the English countryside.

Botanical collecting was close to the heart of the British Empire. Imperial botanists made moving plants around the world into a major industry. When Britannia ruled the waves, the ships of the Royal Navy rearranged the planet's biology. They took eucalyptus trees from Australia across the world, initially to fight malaria by using their prodigious thirst for water to drain swamps. They took potatoes from the Americas to Europe and then Asia, blackberries from the hedgerows of England to Saint Helena, rice from Asia to South America. They also took camels from Arabia to British Columbia and Alice Springs, donkeys from Ireland to South Africa, toads to Bermuda and almost everything to Australia.

Moving plants formed the backdrop to many colonial escapades. Captain William Bligh sailed to Tahiti in the South Pacific to bring breadfruit trees to the West Indies as a cheap food source for plantation slaves, who had themselves been shipped in from Africa. His first attempt in 1789 aboard HMS *Bounty* ended in an infamous mutiny. Bligh was cast adrift at sea, but made it home and on a second journey in 1793 successfully delivered plant samples to Jamaica. Breadfruit remains a favourite food in the Caribbean today.

The British were not alone. The early German explorer Alexander von Humboldt shipped no fewer than 6,000 species from South America to Europe.[18] Portuguese adventurer Ferdinand Magellan funded the debts he incurred on the first circumnavigation of the Earth in the 1520s by selling cloves he collected from the Moluccas islands of modern-day Indonesia. The Dutch later gained a global monopoly on spices grown on these islands. It was broken in 1753, when horticulturalist and missionary Pierre Poivre, the French governor of Mauritius in the Indian Ocean, bagged cuttings during a series of clandestine visits and planted them in his botanical garden at Pamplemousses.

The infrastructure of colonial plant introductions grew extensive during the 19th century. Kew Gardens in south-west London became a scientific hothouse and holding ground for dispersing plants across the British Empire. Here in the mid-19th century, Joseph Hooker masterminded the planting of Ascension Island. But he had bigger fish to fry. He hatched schemes to smuggle from the forests of South America both cinchona, the source of quinine, and then, most famously and profitably, wild rubber – to create huge plantations in India and the Far East. Oil palm from West Africa, tobacco from the Caribbean, tea from China, and sugar cane from New Guinea, all passed through Kew en route to commercial exploitation in lands distant from their homes.

Kew had outposts in Cape Town, Ceylon, Malaya, Australia, New Zealand, Hong Kong and elsewhere, while the Dutch botanic garden in Leiden linked up with tropical collections assembled in the Dutch East Indies, such as that at Bogor in Java. By no means all the translocations succeeded. While the British rubber project was a huge success in Malaya, its cinchona plantation in India lost out to a Dutch rival. These gardens also became frequent sources for escapes into the bush, as migrant species leapt fences from botanic gardens or plantations and took the chance of freedom offered by the imperial adventure.

The legacy of these escapes lingers. Philip Hulme of Lincoln University, New Zealand checked out the likely origins of 34 plant species on a list of the top 100 invasive species compiled by the IUCN. He found that nineteen of them had initially escaped from botanical gardens.[19] Black wattle, a species of acacia, began its spread across an estimated 2.5 million hectares of South Africa this way. The trumpet tree escaped the Germany-run Limbe Botanic Garden in Cameroon for the forests of nearby Mount Cameroon, abetted by birds and bats that carried its fruit. Water hyacinth was dumped over the fence at Bogor Gardens on Java and never looked back.

Domesticated animals also have a long history of riding the colonial ships and then escaping into the wild. With four legs to carry them over the horizon, they have often been spectacularly successful colonists. Cattle and sheep were introduced widely by European powers to turn their new colonial possessions into sources of food for colonialists. As early as the mid-16th century, there were huge migrating herds of feral sheep from Mexico as far south as Santiago in Chile. By the following century, cattle grazed North America from New England to Florida. By then too, horses had been shipped out to the Americas, journeying in the first ships, including those of Christopher Columbus and Hernan Cortes. Many soon ran wild.

The mid-19th century saw a craze for taking a much wider variety of European species to exotic lands, as well as bringing exotics home. Many schemes were run by Acclimatization Societies set up from New York to New Zealand to London – where, in the early days, members laid plans for some bizarre introductions to the green fields of England. On the agenda were wildebeest from Africa, bison from America and armadillos from South America. The British Acclimatization Society's leader, surgeon and zoologist Frank Buckland, who reputedly ate everything from moles to bluebottles, drew up a menu for the inaugural dinner. It included birds' nest soup from Java; *nerfs de daim*, a soup made from the sinews of Axis deer from Sri Lanka; kangaroo steamer from Australia; curried chicken from Siam; sea cucumbers from China; Syrian pig; Honduran turkey; Canadian goose; sweet potatoes from Algeria; dried bananas from Réunion; and guava jelly from Dominica.[20] But despite all the talk, and all the eating, introductions to England were few and the society was wound up in 1865. Its most notable legacy was the colonies of red-necked wallabies that survived on Midlands country estates into the 21st century, before succumbing to one too many cold winters.[21]

Exports worked better, not least because early colonists often pined for familiar landscapes with familiar flora and fauna, wishing to hunt familiar quarry and to spend their evenings dining on familiar meals. In the Cape area of South Africa, European mountaineers are said to have taken packets of European seeds with them on their climbs to sprinkle as they walked on Table Mountain and elsewhere, with the hope of brightening up what they saw as dull scenery. Thomas Austin, the man who introduced the European rabbit to Australia, was a member of the Acclimatization Society of Victoria.

So far as the colonists could see, there was plenty of room for introductions. New Zealand, as one chronicle put it, had 'few wild animals, few flowers, streams almost destitute of fish … with shy songbirds and few game birds and no quadrupeds but lizards'.[22] So in came California quail, Egyptian geese, Australian possums, Tasmanian trout, Arabian camels, American Chinook salmon and English red deer, house sparrows, stoats and ferrets. While many died off, others were successfully released into the wild or soon escaped their patrons. The common brushtail possum, shipped from Australia primarily for its fur, has grown bigger than its cousin back home, and now outnumbers humans seven-to-one.

The British took a small Indian mongoose called *Herpestes javanicus* to many places, employing its lightning speed to kill snakes and chase down rats. It had fun in the Caribbean, eating chickens, devouring innumerable turtle eggs, and attacking lambs, dogs and cats. The French were up to similar tricks in their empire, at one point releasing European hedgehogs in Martinique to eat the snakes. Rabbits introduced to one corner of Chile smartly hopped across the Andes onto the Argentine pampas.

But the acclimatization movement was at its wildest in the United States. As early as 1846, the Natural History Society of America was releasing European songbirds into the skies above Brooklyn. It rustled up 58 pairs of English house sparrows to combat a plague of worms in Greenwood cemetery on Long

Island. Chapters of the Society were at work from Oregon to Massachusetts and from California to Hawaii. A country club in San Francisco was founded to bring brown trout from Europe to the state's streams. In Cincinnati, after housing them in a European-style mansion for a few weeks, the Society released robins, wagtails, skylarks, nightingales, corncrakes, blackbirds, dippers and tits. Most were never seen again.[23]

In New York, the high priest of the movement was Bronx pharmacist Eugene Schieffelin. Sometime in the 1870s, he conceived the idea of bringing to New York every European bird mentioned in the works of William Shakespeare: some 600 in all. Skylarks and finches, robins and blackbirds, sparrows and tit-mice, all had their moment in the New World thanks to the eccentric Schieffelin.[23] Most failed to survive for long. But Schieffelin struck lucky when in 1890 and 1891, he released 40 pairs of European starlings (*Sturnus vulgaris*) in Central Park. His inspiration was a brief reference to the bird in an interchange between Hotspur and the Earl of Worcester in *Henry IV Part 1*.

Three years later, the first nests were spotted under the eaves of the American Museum of Natural History adjacent to the park. By the following year, they were on Long Island and from then on there was no stopping them. Like humans, starlings had taken a liking to the New World. Today, the starling is North America's most numerous bird, with an estimated 200 million on the wing. There are almost as many as humans. Most are descended from the Central Park releases. Their impact on native bluebirds, woodpeckers and others that they pushed from suitable nesting sites is still debated. Meanwhile, Central Park followed up with a garden devoted to the plants mentioned by Shakespeare. It lives on, and there are others scattered across the US and elsewhere, an echo of the time when acclimatization was all the rage.

*

Another Victorian whim that has taken off round the world is a big-leafed water weed whose brilliant purple and violet flowers and dark leathery leaves float prettily across the backwaters of the Amazon and its tributaries: water hyacinth (*Eichhornia crassipes*). It delighted local rubber barons at the height of the rubber boom there in the 1880s. These 19th-century equivalents of hedge fund bosses and dot-com billionaires took to growing it in the ponds of their mansions in the Peruvian jungle city of Iquitos. From there, someone shipped some out of the Amazon. Perhaps it was the notorious Julio César Arana, who ran chain gangs harvesting the sap of rubber trees across an area the size of Belgium, and who also had a home at Biarritz in the south of France.

The truth is murky, but before long, the pretty weed was the fashionable foliage in ornamental ponds across the imperial globe. It was among the many plants on display in 1884 at the New Orleans Cotton Exposition, and by 1896 it had escaped in sufficient quantities to become a navigational hazard in the waterways of Florida.[24]

Water hyacinth is, according to Paul Woomer, an agricultural scientist formerly at the University of Nairobi who has written widely on its progress, 'one of the fastest growing and most aggressive plants on Earth'. It usually spreads by sending out short runner stems that produce daughter plants. But it also produces seeds that can survive in soils or lake sediments for up to fifteen years. Theoretically, a single plant can produce 140 million offspring in a year. An 'infestation' typically doubles in size within a week.[25]

The spectacularly fecund colonist has spread to more than 50 countries, from the backwaters of Bengal to the rivers of China, from Fiji to Australia, from the Louisiana bayous to the Mekong delta. It follows humans, feeding hungrily on nutrient pollution from sewage and farm runoff. It turns up in lakes and wetlands, estuaries and ponds, slow-flowing rivers and irrigation canals. It even has beachheads in Corsica and Spain – not helped, as one exasperated researcher told me in 2013, by the fact that it is on sale to gardeners at supermarkets across Spain. The IUCN lists water hyacinth among the world's ten worst weeds. The charge sheet includes blocking waterways, destroying fisheries and harbouring diseases.[26]

In some places, we know just when it broke loose and transferred from pretty pond adornment to waterways menace. In the early years of the 20th century, the botanic garden at Bogor in Java began dumping unwanted water weed over the fence into the Ciliwung River.[27] Water hyacinth is now everywhere in Indonesia. It was first sighted in the wild in South Africa in 1910, reached Zimbabwe in 1937, Zambia and Malawi in the 1960s, Kenya by the 1980s, and entered West Africa via Cameroon at the turn of the millennium. Almost no large water body in Africa remains untouched, although the extent of infestation varies greatly.

In East Africa, the weed entered Lake Victoria, the world's largest tropical lake, after floating down the River Kagera in Rwanda. One story is that it escaped from ponds in gardens once tended by Belgian settlers there. During the 1990s, the Kagera carried a constant stream of hyacinth clumps into the lake. In 1994, they were accompanied by – and perhaps nourished by – the rotting bodies of tens of thousands of victims of Rwanda's genocide. From the lake's western shores, the weed spread across open waters and round shallow, muddy shorelines. By 1998, when I visited, it had covered four-fifths of the Ugandan shore and ships took five hours to push through the weed to dock at the Kenyan lake port of Kisumu.

Fishers at Kusa Bay, south of Kisumu, showed me mats of weed up to two metres thick that had filled their bay and destroyed their livelihoods. More than 50 fishing boats were marooned on the weed-infested shore, the fish warehouses were permanently locked and the old shorefront hotel derelict. There was hyacinth as far out into the lake as the eye could see. It harboured snakes, crocodiles, hippos, snails that carried bilharzia, mosquitoes that transmitted malaria, and clods of faeces that might have contained cholera. One small village on the lakeshore had lost five people to crocodiles and three to hippos in the previous year.

In Uganda, at the hydroelectric plant at Owen Falls, four boats fitted with rakes and conveyor belts struggled to keep the weed clear of the turbines. At Port Bell outside the capital, Kampala, a floating harvester, purpose-built in Britain, sank under the weight of weed on its first day in the lake. In any case, the IUCN's Geoffrey Howard told me, harvesters 'just create space for the plants to grow into'. Herbicides had failed. In despair, scientists had deployed two species of South American weevil (*Neochetina eichhorniae* and *N. bruchi*) that ate the weed's leaves. If they eat enough, the leaves dry out and weevil larvae can tunnel into the stem, eventually killing the weed. Weevils were credited with limiting the spread of smaller infestations in India, Thailand, Florida, Java and Zambia. But when I visited the fisheries research station near Kisumu, where large numbers of weevils had been bred and put into the lake over the previous two years, water hyacinth still grew with alacrity. 'The bugs can only eat so much', one researcher told me.

It seemed hopeless. And then everything changed. Less than a year later, the weed was in rapid retreat, Kisumu's port was open, cargo boats moved with ease, fishermen were back in business. What happened? Some claim it was a success for the weevils.[28] They must have helped. But what had changed dramatically in the intervening months was the quality of the lake's water. It had been flushed clean, and with it the weed had departed.

This lends credence to a theory that the real problem in Lake Victoria was never so much the weed itself as the choking pollution. Over the years during which the weed had taken hold, the lake and the rivers that delivered water to it had filled with sewage from cities such as Kampala and Kisumu. Effluent from sugar factories, paper mills, tanneries and breweries proliferated across the lake's catchment; and silt washed into the lake as forests were chopped down and soils eroded. This pollution had turned a once-clear, well-oxygenated lake into a muddy water body with no oxygen in its bottom layers. Water hyacinth, raised in the fetid swamps of the Amazon, loved it. But then things abruptly changed.

In 1998, there was a major El Niño event in the Pacific Ocean. It caused unusual weather around the world, including massive and unseasonal rains in East Africa. During my visit that year, much of Nairobi had been flooded. Rivers brought torrents of fresh water that cleansed the lake of its accumulated pollution, says Lawrence Kiage, a Kenyan researcher now at Georgia State University.[29] The weed was suddenly starved of the rich soup of nutrients on which it depended.

The story fits, as do subsequent events. There has been no big new El Niño since 1998, and no exceptional rains to flush the lake clean. The pollution is building up again, and in 2013 the hyacinth was advancing once more. The bays and beaches were covered in weed. The weevils, if they were still around, were not keeping it back. Fishermen were again abandoning their boats. It appears that blaming the alien weed – and seeking solutions to the lake's problems by trying to remove it – has been the wrong diagnosis and the wrong treatment. Lake Victoria remains sick. It may never be freed of weeds until its wider ecological problems are solved. That, as I hope to show, is a lesson relevant to many other infestations of alien species.

Casting the weed as an environmental villain has also meant that water managers and conservationists alike have been blind to the idea that water hyacinth might be a resource of some value. It might be used and harvested – even valued as a crop, under the right conditions. The evidence for this heretical thought comes from China, which deliberately introduced it in the 1930s as a way of absorbing heavy metals in water bodies, and as a livestock fodder crop. It continued to do so until the 1970s, when mounting pollution from the fast-growing city of Kunming caused the weed to take over Dianchi Lake, China's sixth-biggest lake. But Chinese scientists didn't give up. In 2010 they began planting water hyacinth in artificial enclosures in the lake to soak up nutrients. They harvested the weed and processed it to produce bio-gas and organic fertilizer for local farms. The result was both an economic benefit and a cleaner lake.

Zhi Wang of the Jiangsu Academy of Agricultural Sciences in Nanjing claims the experiment 'suggests a tremendous potential for the utilization of water hyacinth for nutrient removal'. Maybe Lake Victoria, which is 200 times the size of Dianchi, is too big to achieve similar results. But the Chinese findings do suggest a creative solution that could work in many places 'infested' with the weed. Perhaps we should try this trick more often with other alien species – thinking about their vigorous growth as a potentially valuable resource rather than a threat.[30]

Lake Victoria has another widely touted environmental villain. The Nile perch (*Lates niloticus*) was introduced to the lake by British colonialists in the 1950s, initially for sport fishing.[31] The fish is a giant. It grows up to two metres long and has a prodigious appetite. It has become one of Uganda's most valuable exports. But it is also blamed for the extermination of around half of the 500 species of small bony cichlid fish that once lurked in the muddy lake depths. Cichlids are widely kept in home aquariums. The species include angelfish and oscars. In Lake Victoria, they have been a uniquely dominant presence. Most of the species there evolved in the lake itself and are found nowhere else. They once made up an estimated 80 per cent of the lake's fish population. The famous Harvard ecologist Edward O. Wilson called the loss of so many of these primitive fish 'the most catastrophic extinction episode of recent history'.[32] But some question whether the appetite of the Nile perch is to blame for their demise.

Dirk Verschuren of the University of Ghent suggests that the real damage was done by pollution. The demise of the cichlids coincided with the accumulation of polluting nutrients in the lake that both fed the water hyacinth and exhausted oxygen supplies in the lake bottom. That process was under way before the first perch were dropped into the lake, and intensified subsequently. The deep waters were occasionally devoid of oxygen in the 1960s and regularly so by the 1980s, which is when both water hyacinth and Nile perch started taking over the lake. Fighting for breath and forced to leave their lake-bed lairs, the cichlids became easy meat for the perch – which duly exploded in numbers while eating their way through the indigenous fish stocks, says Verschuren.[33]

There has been a partial revival in cichlid numbers – and some increase in species surveyed, too – since the early 1990s, in the aftermath of the lake's flushing. Much remains unclear, but it seems likely that it was the demise of the cichlids that allowed the takeover by the Nile perch – rather than the other way round.[34] The lesson from Lake Victoria is that in two highly publicized cases, the bursts of activity by alien species were made a scapegoat for much wider problems of pollution and environmental decline in the lake's catchment. The aliens took advantage of the environmental crisis, but they did not cause it. Demonizing the alien species obscured this and prevented effective action to halt the problem.

*

Africa has been the victim of poor science and outrageously ill-considered policies on alien species foisted on it from outside. Its environment often seems to be a pawn in wider political games, many of them resulting in policy U-turns.

How else can we explain the changing perceptions of a

thorny evergreen bush from the Mexican desert called mesquite, or *Prosopis*? The bush copes with drought by sinking roots tens of metres to find water. That makes it a good colonist in arid lands. But anyone reading the international literature on the plant will be confused about whether we should love it or loathe it.[35]

The United Nations and many international development agencies once called mesquite a 'wonder tree'. They planted it across the dry famine-prone lands of sub-Saharan Africa from the 1970s right up to the late 1990s. They promised that, by tapping water from deep underground and surviving where other plants could not, it would bring life back to areas hit by spreading deserts in Ethiopia, Sudan and elsewhere. It would give locals an alternative source of firewood, while its nitrogen-fixing properties improved farm soils and its gum treated minor infections. Birds would nest in it. Bees would be attracted to its nectar. Its virtues seemed endless.[36]

One of the bush's most feverish and persistent advocates was the UN Environment Programme (UNEP), which wanted to halt desertification in Africa. As recently as 1997, it said that its mesquite-planting projects in Sudan and elsewhere were 'helping communities protect their environment' by halting the desert's advance, and 'bringing new areas into cultivation'.[37]

But minds changed. A decade later, UNEP was warning against what it called 'mesquite mania'. In a book on invasive plants in Africa, Arne Witt, the Nairobi-based UNEP consultant on invasive species in Africa, said that 'most introduced *Prosopis* species have turned out to be relentlessly aggressive invaders … rampant across Morocco, Algeria, Mali, Burkina Faso, Niger, Chad, Sudan, Somalia, Ethiopia, Kenya and South Africa'. Early benefits had been 'overshadowed by their negative, invasive impacts'.[38] Mesquite had created 'sprawling impenetrable thickets'. Far from halting desertification, it was creating 'new green deserts' by lowering the water table and killing native trees. Far from feeding animals, it had 'replaced grasses and reduced livestock carrying-capacity on traditional pastures'. In places it had forced communities to abandon their land. Governments were launching eradication programmes. In this long diatribe, Witt omitted to mention the role of UNEP in what it now regards as a dreadful mistake.

Some pastoralist communities were in trouble. In 2006, people who lived along the shores of Lake Baringo in northern Kenya claimed that the plant had spread rapidly, consuming their water supplies, ripping their flesh with its thorns and killing their goats. They sued the Kenyan government, which had planted the mesquite at the behest of the UN Food and Agriculture Organization. At one point the community paraded a toothless goat through the courtroom, to show how sugary fibres of the plant's seed pods caused gum disease.[39]

Clearly this was not good. But the truth is more nuanced. In a study around Lake Baringo, Esther Mwangi of Harvard and Brent Swallow, now at the University of Alberta, found that the invader had created winners as well as losers. The winners were typically those who harvested and sold wood for construction, mesquite seedpods for feeding penned livestock, and bark for making rope. The pods had even become an export crop. Others sold honey made by bees attracted to the bush. Most losers were pastoralists who had lost grazing land.[40] A British government study dismissed the new fad for trying to eradicate mesquite as bone-headed. 'Eradication is not only impossible but also unnecessary', it said.[41]

It is a shame that UN agencies, conservationists and others cannot promote such balanced thinking rather than engaging in blind rhetoric that only begs the question of when they might change their minds again.

3

All at Sea

They called it the blob that ate the Black Sea. Sometime in 1982, a previously harmless New England jellyfish – a plankton-grazer the size of your hand, known to marine biologists as *Mnemiopsis leidyi* – got into the ballast tanks of a ship delivering cargo to the US east coast. A few weeks later, the jellyfish was discharged into the Black Sea, probably near the Ukrainian port of Odessa.

Sometimes a newcomer in a strange land strikes it lucky. For a while, the tentacled semi-transparent jellyfish lurked unremarked, but in 1988 it went on manoeuvres. *Mnemiopsis leidyi* may not sting, but it packs a punch. A self-fertilizing hermaphrodite, it bred as fast as it ate, reaching maturity within two weeks and producing 8,000 eggs a day. With no predators around to eat those eggs, its rate of reproduction was staggering. It found a ready food supply and began munching its way through the sea's plankton and crustaceans and the eggs and larvae of local fish.[1]

Soon its gelatinous mass was breaking records. A snorkelling marine biologist from the Ukraine, Yu Zaitsev, counted 500 of the blobs floating in a single cubic metre of water in Odessa Bay. There was almost more jellyfish than water. It was clogging fishing nets. Before long it was the only thing caught in the nets. Nothing could live with it, especially not the local jellyfish, *Aurelia aurita*. Catches of anchovy, mackerel and other fish crashed. Dolphins disappeared. By 1990, biologists estimated the total weight of the jellyfish in the Black Sea at 900 million tonnes. If true, that was 95 per cent of the animal biomass in the Sea, and equal to ten times the annual fish catch in all the world's oceans.[2]

It couldn't last. With little left to eat, *Mnemiopsis* began to go hungry. And before too long, its nemesis showed up. Another shipload of ballast water dumped into the sea in 1997 brought another American jellyfish, *Beroe ovate*, which eats nothing but its cousin. Since then, *Mnemiopsis* has been in modest decline in the Black Sea, reaching an uneasy balance with its food supply and the new predator. But it has been moving on. In 1996, fishers from Turkmenistan reported finding strange jellyfish in their nets in the Caspian Sea. *Mnemiopsis* had hitched a ride aboard ships travelling the Volga-Don ship canal between the two seas. By 2000, its insatiable appetite had cut fish stocks in the Caspian by more than 50 per cent, outcompeting local seals and beluga sturgeon for a local delicacy, the kilka fish. Since 2006, *Mnemiopsis* has turned up, less problematically so far, in the North and Baltic Seas. And it is spreading through the eastern Mediterranean, clogging the intake pipes of Israeli desalination plants.[3] The blob is not done yet.

That is the conventional story. As on land, when we are looking at the spread of alien species at sea, the temptation to demonize them is strong. But again it usually ignores the context – the way our own activities typically pave the way for the invader. Blaming a chance arrival in American ballast water for the collapse of Black Sea fisheries may be convenient, but it is not good enough. Such invasions do not happen out of nothing. The Black Sea was a fetid haven for the blob. Europe's most polluted sea was a blocked toilet bowl. It collected raw sewage from about 170 million people in thirteen countries floating down from rivers such as the Danube and Dnieper. Its catchment extended from southern Germany almost to Moscow, but it had only a tiny outlet into the Mediterranean through the 800-metre-wide straits of the Bosphorus. A full exchange of water with the Mediterranean took about a thousand years. Because the sea could not flush itself clean, the poison built up. The sea's depths had long been devoid of oxygen. But about 50 years ago, the anoxic waters reached the surface and spread across the sea's large, shallow north-west shelf, which had previously been its most fertile zone.

The final straw that let in the blob was the death of huge beds of red *Phyllophora* algae on the north-west shelf, according to Lawrence Mee, a Scottish marine biochemist who ran the UN's effort to save the sea at the height of the blob's reign. 'In the 1950s, these underwater meadows covered an area the size of the Netherlands. They were the lungs of the Black Sea', he said. They emitted oxygen as well as providing nourishment and shelter for the sea's large stocks of anchovy, turbot, mackerel, flounder and sturgeon. 'As the meadows died', said Zaitsev, 'there was a colossal death of bottom-dwelling organisms, such as sponges, sea anemones, shrimps and crabs, and a domino effect right through the marine ecosystem.'[4] Just like water hyacinth's invasions of Lake Victoria, the blob did not so much eat the Black Sea as move in when the sea had died. A healthy sea would in all probability have fought off the invader.

*

That said, ships' ballast water certainly provides migrating species with an opportunity to cause mischief. It is probably the most important means by which alien species spread around the world today. There is a strong case for avoiding disruption by preventing discharges of ballast water containing organisms from distant waters.

Most cargo ships and tankers need to add weight when empty, in order to avoid capsizing. So they take on ballast, which they dump when they fill up with cargo. Once, this ballast was rocks or soil. Ships crossing the Atlantic routinely grabbed chunks of the geology of south-west England before they left. Days later, they deposited it – with Cornish plants, worms and other organisms – onto the shorelines of the US and Canada. But, ever since the introduction of steel-hulled ships in the late 19th century, seawater has been the ballast of choice.

We are talking about a lot of water. A large ship can carry 60,000 tonnes of ballast. There are tens of thousands of vessels at sea. Altogether, an estimated 7 billion tonnes of ballast water is shipped round the world each year, usually discharged into foreign coastal waters. Often it is rich in seeds, spores, plankton, bacteria, and the eggs and larvae of larger creatures. Marine biologists estimate that at any one time, the ballast water travelling the world's oceans contains 7,000 different species. Some of that biological cargo will be ready to take advantage of whatever ecosystems they encounter on arrival.

Soon after the jellyfish from Maine made its fateful journey to the Black Sea in the early 1980s, the Central Asian zebra mussel (*Dreissena polymorpha*) hitched a ride in the opposite direction. It ended up discharged into the Great Lakes. We will look at the consequences in the next chapter. Around a decade later, an unknown ship, probably from the Bay of Bengal, discharged ballast water into the coastal waters of Peru. That water contained a strain of cholera that was absorbed by local shellfish. People ate the shellfish and the disease spread in excreta and waterways, killing 12,000 people across Latin America in the succeeding months.[5]

Ballast water often delivers dinoflagellates, forms of algae that cause toxic 'red tides' in polluted waters around the globe. It is how Chinese mitten crabs reached Europe, how Asian kelp made it to southern Australia, and how Mediterranean mussels came to carpet the coast of South Africa. It is also the biggest reason why the Mediterranean itself, the world's busiest sea, contains an estimated 900 marine species that biologists regard as alien. A new one arrives in the Med every nine days, according to the UN Environment Programme. Most soon die off. Most of the rest are harmless. But the risk of a new zebra mussel or *Mnemiopsis* jellyfish or disease infestation is ever-present.

An effective invasion is most likely if the journey is not so long that the alien critter dies en route, and not too short that it may already be present locally. Invasions are also more likely if the starting point is in warm water, which usually contains more species, and if the water at both ports is of similar temperature and salinity. That flags up busy tropical ports handling long-distance cargo, says Bernd Blasius of the University of Oldenburg in Germany. After mapping more than 3 million ship journeys, and assessing the species migrations possibility presented by each, he calculated that just twenty major ports carry 39 per cent of the total risk of invasions. They include San Francisco Bay, New York, Los Angeles, Hong Kong, Singapore and Durban in South Africa, along with the Suez and Panama canals. The journeys most likely to deliver aliens are between 8,000 and 10,000 kilometres in length – roughly the distance between busy Asian ports such as Shanghai and Singapore and the west coast of the US.[6] The prediction matches data from ports that have numerous invaders, such as San Francisco Bay, where the majority of inhabitants are alien species, whose names often tell their own story. Scoop water from the bay and you may find Chinese mitten crabs, New Zealand sea slugs or Japanese gobies – all brought by ballast water. We will look at the state of the bay further in the next chapter.

So, many ask, why the delay in addressing the threat? It is a good question. A treaty agreed at the United Nations' International Maritime Organization (IMO) in 2004 required most international cargo ships and tankers to install disinfection kit to kill off biological stowaways.[7] But, as of late 2014, it had not been ratified by sufficient nations to come into force. The technology is available. The IMO has certified more than twenty commercial treatment systems involving various combinations of filtration, irradiation, ozone dosing, heat, electrolysis and biocides. The hold-up seems to be the cost. At up to $500,000 in equipment for each of the biggest vessels, we are talking about a billion dollars or more.

But there should be another concern, says one prominent

marine biologist, speaking on condition of anonymity. Some methods of cleaning up ballast water could actually increase the risks – in particular from bacteria. The warm and quiet conditions inside ballast tanks can act as incubators for colonies of bacteria. And some treatment systems could make matters worse by killing off the things that eat the bacteria, like copepods, while leaving the bacteria unscathed. Moreover, the dead plankton floating around in the tanks after the treatment could provide a valuable source of food for the bacteria.

'You could end up with 10 or 100 times more bacteria, as a result of the wrong sort of treatment', he told me. Sitting in the tanks at higher concentrations, the bacteria would also be more likely to turn into new strains through genetic exchanges. 'You could be creating new diseases', he said, especially if you treat at the start of a voyage but not again before discharge. With no present IMO requirements to measure bacteria in the tanks during or at the end of a voyage, it is a disaster waiting to happen, he believes.

It may not be so bad. Bacteria and viruses already travel the world in their billions in ballast water. Most of these microbes may be present so widely in the oceans that moving them around makes little difference to the threat they pose. But Fred Dobbs, a marine microbiologist at Old Dominion University in Norfolk, Virginia, says: 'the bacterial and viral diversity in ballast water is absolutely unknown.' And the story of cholera reaching Peru suggests a significant, if infrequent, risk.[8]

*

You don't need ships to create an alien invasion at sea. During the autumn of 2013, I joined a tour of the Monaco Oceanographic Museum. The museum was established by the late French marine adventurer and conservationist Jacques Cousteau on the shores of the Mediterranean. After seeing all the aquariums, with their magnificent specimens of marine life, we descended into the vaults where they kept the sick fish, prepared displays and did the museum's plumbing. My guide, an in-house scientist, explained how they took water for the tanks from the sea just outside the walls of the museum. 'And do you discharge the waste into the sea too?' I asked. He looked a little uneasy. 'Yes. Did you have anything in mind?' Well, I did. '*Caulerpa*', I replied.

I had been in Monaco twenty years before to interview marine biologist Alexandre Meinesz, from the nearby University of Nice. Back then, he publicly accused Cousteau's aquarium of causing ecological meltdown in the Med. One of Meinesz's students had gone diving and spotted a small patch of bright green weed on the sea bed right outside the museum. It resembled Astroturf and was close to the pipe used to discharge used water from the aquariums. The weed was previously unknown in the Mediterranean. It turned out to be *Caulerpa taxifolia*, an algae native to the Indian Ocean. It had been imported to Europe a decade earlier for use in aquariums at the Wilhelma Zoo in Stuttgart, Germany, from where Cousteau took some to brighten up his displays in Monaco. Evidently, some had escaped.[9]

Nobody removed the patch of weed on the sea bed. For five years, it grew only slowly. But after 1989, it rapidly spread down the coast, blanketing 300 kilometres of the Riviera from Toulon in France to Imperia in Italy. Soon, it turned up in the Balearic Islands and as far south as Tunisia in North Africa and as far east as Croatia. Meinesz raised the alarm. It was killing the underwater meadows of sea grasses, where hundreds of fish species spawned and fed, he said. It had mutated, growing to more than twice its normal size and generating a higher dose of caulerpicin, a toxin that kills fish, sea urchins and anything else that eats it. 'We could be seeing the beginning of an ecological catastrophe', he told me one sunny morning at his campus office.

The panic grew. Marine biologists around the world went on alert for the mutant algae. In 2000, it turned up in a lagoon in San Diego, California, probably after release from an aquarium close by. A scratch team of snorkelling scientists was down there within days, ripping it up with their bare hands and applying chemicals. They called themselves a 'self-appointed ad hoc management team of exceptionally committed local managers'. And they got $6 million from a local corporation.[10] Meinesz, meanwhile, wrote a best-selling book, *Killer Algae*. *Caulerpa* had 'devastated the Mediterranean ecosystem [and] defeated the French navy', he wrote. If only Cousteau's crew in Monaco had done the same as the San Diego snorkellers, the sea would have been saved, he suggested. But the aquarium was still denying responsibility.[11]

Fast forward to 2013, and my guide at the museum was still not prepared to admit his employers' responsibility. 'It may not have come from here', he insisted. 'There are other places in the Mediterranean where they use *Caulerpa*, and anyhow it is disappearing now', he said. The tour continued. But I was intrigued. Not much had been written about the 'killer algae' lately, so I checked the literature. Had Meinesz's dire predictions come true? No, they had not. *Caulerpa* was disappearing as fast as it had arrived, and had entirely gone from large areas, as Meinesz agreed when I approached him. The *Caulerpa* binge in the Mediterranean was largely over. The algae were dying in many areas. The demise had been going on along the Riviera almost since Meinesz's book hit the stores. In 2003, there was less than a tenth as much *Caulerpa* as Meinesz had recorded a decade before. Published papers have since reported its disappearance from the shores of Italy and Croatia.[12]

What had been going on? When writing up my interview with Meinesz in 1992, I had called John Chisholm, an Australian marine biologist at the European Oceanological Observatory in Nice. He told me that he thought the noxious weed was simply taking advantage of a Mediterranean ecosystem badly damaged by pollution. '*Caulerpa* grows in sediments rich in organic pollution – from sewage outfalls and so on', he told me. The sea grasses that Meinesz said had been killed off by *Caulerpa* were already in a bad state before the alien weed came down the pipe from the museum. The weed was not a killer but an opportunist. Its spread was 'the long-term effect of pollution over two or three decades in the Cote d'Azur'. If the pollution receded, so, in all probability, would the infestation.

I must admit that at the time I didn't entirely believe him. But it turned out Chisholm was right. In fact, says Tim Glasby of the Port Stephens Fisheries Institute in New South Wales, who had seen a similar boom and bust with *Caulerpa* in Australia, its arrival could be seen as the first stage of ecological recovery along the Riviera, rather than decline. The new arrival had not pushed out rich sea meadows, but colonized bare rock and sediment that the meadows had vacated. It was saviour rather than villain. It has turned out that many *Caulerpa* beds in the Mediterranean had more marine life, and more species, than the former sea meadows. Clams did especially well; cockles hid from predators in its green embrace. The weed from the Indian Ocean liked sewage pollution so much that it was an effective means of treating the stuff. It might, theoretically at least, eventually help the return of sea grasses. It began to look like Jacques Cousteau's poor aquarium hygiene three decades ago had done the Mediterranean a good turn. And that the 'killer' algae's recent decline might be a bigger problem for marine life than its arrival.

To me this was a fascinating story, at least as interesting as the original upsurge in the weed – and arguably much more revealing. Yet marine biologists seemed much less eager to report and analyse the retreat of *Caulerpa* than its spectacular arrival. Published papers were rare and the entry for *Caulerpa* on the supposedly authoritative Global Invasive Species Database of the IUCN had not been updated for almost a decade. In late 2013, it said nothing about the weed's demise. It claimed that the weed still covered more than 12,000 hectares along the Riviera, something that had not been true for many years. And it continued to insist that *Caulerpa* 'excludes almost all marine life', which is demonstrably untrue.[13]

The stories of *Caulerpa*, *Mnemiopsis* and many others show how we instinctively blame aliens for ecological problems that may have a lot more to do with our own treatment of nature. Like xenophobes in human affairs, conservationists too often think the worst of the aliens. But if we must think in terms of good and bad, then aliens can be good, too. Successful aliens are usually just taking advantage of what they find. Indeed, on occasions, they are, if we give them a chance, part of the solution rather than the problem. *Caulerpa* moved onto bare rocks after sea meadows had been vanquished by pollution, creating not an ecological desert but a rich nursery for marine life.

*

Clearly the oceans are a good medium for species to travel the world. Migrant organisms don't need a ballast tank, though that clearly makes it easier. Anything capable of staying afloat and capturing food can travel anywhere the currents will take it. But equally, in the oceans it is hard to say who is alien and who is not. Indeed, the very idea of species having a unique home to which they are uniquely adapted makes even less sense than it does on land. Many marine organisms cross the oceans as part of their natural life cycles. And these migrants often take unsuspecting species along for the ride. The oceans are alive with tree branches, animal carcases and other organic detritus washed down rivers. The larger items will often be inhabited by other species that either live on their floating home or join en route.

These days there is more junk in the oceans to provide lifts for would-be migrants. Human activity on land, like cutting down forests, creates potential seafarers, and we also provide containers. The oceans are full of plastic: drinks bottles and abandoned fishing nets; six-pack rings and flip-flops. (On remote Ascension Island, Boy Scouts regularly collect flip-flops from their South Atlantic shoreline and enjoy trying to establish their origin.) Floating plastic is virtually indestructible and so has a longer range than most vegetation. We hear from time to time about how it chokes, smothers and starves unsuspecting turtles, albatrosses and even whales. But plastic detritus also carries hitch-hikers, and even victuals for their journey. A water bottle might supply freshwater; a food container could yield scraps of sustenance.

David Barnes, a zoologist at the British Antarctic Survey, scavenged remote shorelines from Spitzbergen in the Arctic to Signy Island off Antarctica, and from Galapagos in the Pacific to Ascension in the Atlantic. He found that more than half the debris was man-made, and much of it was inhabited by worms, ants, barnacles, larvae and even hardy insects. The density was highest in the Southern Ocean, probably because there are fewer shores for the debris to beach. With a year-round circumpolar current, the detritus just keeps on going round.[14] Barnes says that 'the vast amounts of waterborne debris is almost certainly drastically changing opportunities for many marine organisms to travel and thus for exotic invaders to spread'.[15]

Beachings can be spectacular. On 5 June 2012, waves tossed onto the beach of Newport, Oregon, a 20-metre-long piece of jetty. It had crossed the Pacific from Japan, where it had been cut loose by the tsunami fifteen months before. It was a journey of at least 8,000 kilometres, and possibly a great deal more. On Newport beach, demolition teams quickly arrived to break up the wooden structure and dispose of it, for fear that Japanese species were ready to hop ashore and colonize. As they worked, scientists found 90 different species of hitchhiker that had either survived the entire journey or joined up en route. They included an edible Japanese seaweed called wakame that is on the IUCN's worst 100 alien species list; a starfish called the Northern Pacific seastar; and several mussels, crabs and worms.[16]

If humans had not been around to police the arrival, any of

those species might have become Americans. While the jetty is a human artefact, a tree trunk or other large piece of organic debris could have made the same journey and carried the same passengers. The event showed how easy – and over millions of years how frequent – such journeys must be. If the Pacific can be crossed so easily, then nowhere is too remote to be visited by aliens.

One can imagine that a similar kind of journey might have resulted in a land crab making it to Ascension Island. Before humans showed up, Ascension's ferns and seabirds were ruled over by its largest land animal, a land crab (*Johngarthia lagostoma*) found nowhere else. Tens of thousands of the crabs burrow into the island's Green Mountain and go on long expeditions down to the shore to reproduce. But given that the island is some 1,000 kilometres from the nearest speck of land, it is a mystery how the crab got there. It never goes to sea, beyond dipping its claws into the waves while depositing larvae. It most likely arrived when larvae of some land crab from Africa or South America accidentally rode on some floating vegetation. After that, it just evolved to what we see today.[17]

What doesn't travel by sea may fly. Many birds migrate around the globe, following the seasons. Birds that spend part of the year in Britain may winter in South Africa or summer in Siberia. There are flight lines across the Atlantic, from Australia to Japan, from North to South America, and in the case of the Arctic tern, from the Arctic to Antarctica. And birds make mistakes. While I was on Ascension Island, there were reports that one of the 'endemic' Ascension frigatebirds had shown up in the Scottish Hebrides. Perhaps it was similar faulty navigation that resulted in many birds from the Pacific region making it to Hawaii over millions of years, where they found rich pickings and, as we saw in Chapter 1, stuck around, evolving into new species. Without such random migrations, many more places round the world would be lifeless.

Birds do not always travel alone. They take with them, in

their feathers and gullets and feet, seeds and insects and pathogens and much else. Some of these fellow travellers will hop off at the other end and stay put. And where birds don't go, the weather may lend a helping hand. A surprising range of smaller organisms like algae, pollen, fungi and bacteria are routinely wafted into the air, and travel with the wind on long journeys.

In 2005, Ruprecht Jaenicke of the University of Mainz reported that up to a quarter of the dust particles in the atmosphere are biological rather than geological in origin.[18] Some of those particles are far from benign. The atmosphere, like the oceans, is rich in bacteria and nobody knows how far they can travel.[19] The winds have for hundreds of millions of years been a ready mode of long-range transport for a range of nasty pathogens, says Gene Shinn, now at the University of South Florida. He discovered that outbreaks of coral diseases on the reefs of the Caribbean often coincide with dust storms blowing in on trade winds from Africa. On a closer look, he found that during major dust storms in the Sahara, 130 different species of pathogens were blown into the air and wafted to the West Indies. He reckons that tens of millions of tonnes of African dust settles on the Caribbean most years, and much more when African droughts leave the soil bare.[20]

Shinn has identified one West African culprit as a soil fungus called *Aspergillus sydowii*. It was first spotted in the Caribbean in 1983 during an intense drought in Africa. It went on to kill 90 per cent of the region's sea fans, a form of soft coral. The same year saw a big decline in the number of *Diadema* sea urchins, as well as infestations of algae on the reefs where the urchins normally graze. Some researchers contest that the fungus came from Africa, but Shinn believes other diseases have made the same journey. Soybean rusts from Africa have crossed the Atlantic by means unknown and infected plants in the US. One might speculate that microbes picked up by dust storms in China's Gobi desert cross the Pacific, too.

It seems likely that the occasional appearances in Europe of foot and mouth (*Aphthae epizooticae*), a highly contagious viral disease that afflicts cloven-hoofed animals, may have a similar explanation. The disease is rife in the Sahara region of Africa. In 2001, British government vets slaughtered 10 million sheep and cattle during an outbreak, and large areas of the English countryside were cordoned off in an ultimately successful effort to halt the disease's spread. There was no proof, but Dale Griffin of the US Geological Survey said at the time: 'Satellite images show a dust cloud moving [from the Sahara] over the Atlantic and reaching Britain on 13 February 2001. One week later, foot and mouth broke out in the UK. Given that the disease's incubation period is seven days, that is one heck of a coincidence.'[21]

This kind of talk can feed our paranoia about alien species. Security types have started researching such outbreaks, fearful that bioterrorists might take nature's hint.[22] The US government's Department of Homeland Security funds studies into microbes in the air. The *Aspergillus* fungus implicated in killing Caribbean soft coral also causes lung disease in humans. Epidemics of meningitis and asthma, and several crop diseases, have all likely been spread by aerosols of living organisms.[23] The risks are real, but the wider ecological lesson is that microbes have always been in the air. Nature has always been rearranging itself.

We may think of volcanic islands like Ascension as unusual because their recent origin and remoteness mean that their ecosystems are made up of a motley crew of mariner migrants. But much of the world is like that. Nature is constantly in flux and few ecosystems go back very far. Only 10,000 years ago, much of Europe and North America were covered in thick ice. All soil had been scraped away, and with it most forms of life. Everything we see today in these former glaciated zones has either returned or arrived for the first time since the ice retreated.

Looked at from this perspective, the spread of alien species today is merely a continuation of a natural process of the colonization begun when the ice retreated. A broad time horizon shows there is no such thing as a native species. All lodgings are temporary and all ecosystems in a constant flux, the victims of circumstance and geological accident. As the pioneer British ecologist Charles Elton argued, 'were it not for the ice age, we [in Britain] should probably have wonderful mixed forests with wild magnolias and laurels and epiphytic orchids, such as … in China'.[24]

4
Welcome to America

Americans used to love kudzu (*Pueraria lobata*). The vine had bright green leaves and fragrant flowers. It grew really fast. The public was exhorted to plant the 'miracle vine' across the South. In a country blighted by the Dust Bowl, America was buoyed up by the hope that nature could be reborn, and kudzu was the saviour. It was a vegetable embodiment of the New Deal, healing the land and reviving its battered people. Its Asian origins were an irrelevance. America, after all, was a nation of foreigners, founded on the belief that the world could be made afresh. Kudzu was the American way. Yet half a century later, the miracle vine has been recast as 'the vine that ate the South'. It is accused of killing trees, invading croplands, wrecking buildings and downing power lines. What happened? Did the vine change, or did America?

Kudzu is a member of the pea family. It originates from China, where villagers still use its stems to make rope, baskets and paper. Its roots are a staple of traditional medicine and its leaves feed livestock and even humans during famines. But its name is Japanese, and kudzu reached America via Japan, thanks to a diplomat called Thomas Hogg. He was Uncle Sam's consul in Tokyo in the 1870s, and regularly sent local plants home to his brother, who ran a nursery in New York and sold kudzu as an ornamental plant. Thomas Hogg subsequently encouraged Japanese delegates to feature the pretty vine in garden exhibits at international trade exhibitions in the US, including the Centennial Exposition in Philadelphia in 1876, which was attended by 10 million Americans.[1]

Kudzu proved popular. Gardeners called it the 'front porch vine'. *Good Housekeeping* magazine in 1909 lauded the 'delicious fragrance' of its purple flowers, and noted that it would 'flourish where nothing else will grow'. That might have been the first warning sign, but back then it sounded good.[2]

Government agricultural scientists took an interest, too. They recommended kudzu leaves as fodder for livestock. And its ability to grow anywhere attracted the attention of government agents fighting drought. From 1935, the head of the government's new Soil Conservation Service, Hugh Hammond Bennett, promoted it as the lynchpin of Operation Dustbowl, his programme to restore messed-up soils, particularly in the cotton fields of the South. The vine could grow as much as 30 centimetres a day. Its spreading foliage covered the ground and prevented erosion by wind and rain. Its deep roots sought out moisture and contained bacteria that fixed nitrogen from the air, fertilizing soils.

There was a craze for kudzu. It had its own champion on the radio. Georgia farmer Channing Cope broadcast his daily down-home radio show from his front porch, where he had planted kudzu. He formed the Kudzu Club of America and wrote articles praising it in the *Atlanta Constitution*. 'Kudzu is the Lord's indulgent gift to Georgians', Cope declared. It would 'heal' the soils. 'If we will feed the land, it will feed us.' Cope, the king of kudzu, was declared Georgia Conservation Man of the Year. Americans may shudder now at how kudzu creeps across the land. But back then it was seen as capable of restoring both the environment and the economy of the South. From 1935 to the early 1950s, government nurseries grew 100 million seedlings. Roadsides and railway embankments were seeded, and farmers were paid to plant the Japanese vine on their land. Altogether more than 800,000 hectares were covered.

Its abilities seemed endless. When botanists tried to find something that would grow on land at Copperhill in the Appalachian Mountains of Tennessee, which had been poisoned by acid emissions from copper smelters, the only thing that would do the job was kudzu. Where it grew, it sucked the poison from the soils. At the height of Cold-War paranoia in the 1960s, radiation ecologists from the Oak Ridge National Laboratory recommended that kudzu should be part of the military arsenal, ready to be deployed in detoxifying the land should a war with the Russians result in 'nuclear devastation'.[3]

Times changed, however. As early as 1953, kudzu's potential to grow where it wasn't wanted had been noted. In the 1970s, Bennett's successors at the Soil Conservation Service quietly removed it from a list of plants approved for erosion control. By 1997, kudzu became officially part of the problem rather than the solution, when it was put on the Federal Noxious Weed list. It was from now on regarded as a malign and alien competitor to native shrubs, trees and crops. A current government fact sheet says that the former miracle vine 'kills or degrades other plants by smothering them under a solid blanket of leaves, by girding woody stems and tree trunks and by breaking branches or uprooting entire trees and shrubs through the sheer force of its weight'. Kudzu will grow almost any place where sunlight is abundant to drive its rapid photosynthesis. It climbs telegraph poles, strangles trees, buries hedgerows and takes over abandoned buildings. Its fecundity and vigour are now officially a burden.[4] It covers some 3 million hectares across the South, and extends its grip by about 40,000 hectares annually.

The vine hasn't changed. It is still revered in Japan. What has changed in America is the land and people's expectations of the land. Kudzu's foliage is no longer needed to feed grazing farm animals, which now live in feedlots. The pastures are abandoned. No longer kept in check by grazing, kudzu now grows where it is not wanted, spreading unchecked almost anywhere south of the Mason–Dixon line. It is the enemy. The pastures are being turned into woodland, where kudzu is a problem.

There seems to me, as an outsider, to be something cultural

at work here too. Kudzu's incontinent growth, extending its roots underground to form new vines, seems to fit an American image of the Deep South as somehow depraved and unruly. In 1999, *Time* magazine listed the introduction of kudzu to the United States as among the hundred worst ideas of the 20th century, alongside asbestos, DDT, telemarketing, prohibition and *The Jerry Springer Show*.

Kudzu has become inconvenient for farmers and landowners, for sure. But is it bad for nature? The claim is often made, but when I searched for information about how much ecological harm it does, I found that nobody seemed to know. The science has not been done, says Irwin Forseth of the University of Maryland. 'Despite widespread anecdotal statements, little quantitative information is available regarding the ecological effects of kudzu.' He did not deny that there were effects. But amid all the hysteria, nobody has done the research.[5]

Could there be redemption for kudzu? Derek Alderman of the University of Tennessee-Knoxville thinks so. The vine may no longer be needed for fodder, but it has other economic values. There are niche markets as a herbal medicine, including a possible treatment for alcoholism, and for making upmarket paper, baskets, jewellery, jams and even soap. Such uses are widespread in Japan and China.[6] But for now, all that is drowned out. The mythology is spreading faster than the weed. Kudzu is a symbol as much as anything, these days. When I searched for it on the *New York Times* website, kudzu as metaphor had swamped stories about the vine itself. A composer was creating 'the kudzu of Jewish music'; someone's bookshelf had 'shaggy dog plotlines sprouting everywhere, like kudzu'; Grace Hightower De Niro spoke with a 'southern drawl peeking beneath her syllables like a branch under kudzu'; a front rower at a fashion parade was accused of hooking her foot around her ankles 'kudzu-like'.[7]

And there was a story about an insect, the bean plataspid or stink bug. It likes to eat kudzu leaves, and is popularly known as the kudzu beetle. It arrived, like its namesake, from Japan, and is spreading out from Georgia like – you guessed it – kudzu.[8]

*

Kudzu was far from alone among alien plants in winning US government approval in the early 20th century. The Department of Agriculture for a long time had an Office of Seed and Plant Introduction, dedicated to bringing in foreigners, rather than keeping them out. In 1943, David Fairchild, its long-time head and the US's plant explorer-in-chief, boasted that his staff had achieved almost 200,000 introductions.[9] And kudzu is by no means the only alien introduction to go from wonder plant to botanical pariah.

The story of tamarisk (*Tamarix ramosissima*), also known as salt cedar, is most interesting. If kudzu somehow sums up the Deep South, then the story of tamarisk says a lot about the West. Like mesquite in Africa, the thorny tamarisk bush has gone from being a bulwark against advancing deserts to a prime cause of their advance – from a finder of hidden water to a water-guzzling monster that is drying up rivers. For some decades, the people of the American West have been removing tamarisk bushes from river banks and desert margins, almost as a civic duty. It is public enemy number one in the arid West. But there is growing evidence that it may be a victim of trumped-up charges.

Tamarisk is an old-world staple. It is ever present in the Middle East, and turns up from the Atlantic shores of south-west England to Korea, as well as across much of Africa. Its greyish-green leaves first appeared at the Harvard Botanic Garden in 1818. It was bought and sold in the eastern US as an ornamental plant, before government scientists and US Army engineers decided it was a good way to prevent soil erosion out West. Department of Agriculture researcher Mark Carleton, a cheerleader for tamarisk, boasted in *Science* magazine in 1914 that 'there appears to be no limit in dryness of the soil ... beyond which this plant will not survive'.[10] The plant sank long roots that both found water and bound the soil. It didn't mind drought, fire or salt. Its wood was valued, and it had influential friends. In about 1920, the famed ecologist Aldo Leopold installed tamarisk in front of his house, which backed onto the Rio Grande in Albuquerque, New Mexico.

Tamarisk was widely planted, but also spread on its own. It loved river bottoms and kept intruding on iconic American landscapes: round the alkaline springs of Death Valley, for instance, and in the Grand Canyon. It replaced much-loved cottonwood (*Populus deltoides*) and willow and began to be blamed for drying up rivers and emptying underground water reserves. The origins of its abrupt recasting lie in the perennially murky world of water politics in the arid West, says Matthew Chew of Arizona State University.[11]

The story – a botanical version of the great water movie *Chinatown* – goes like this. In the 1930s, mining companies in Arizona and New Mexico were desperate for water. But the rights to water from the local rivers were mostly already allocated to farmers. Somehow, they had to find 'new' water. Or, since that was unlikely, they needed to appropriate it from some other use. Someone noted that tamarisk had a reputation for consuming large amounts of water. So, the mining moguls conceived the idea of proposing to remove the tamarisk in river valleys where they wanted to tap for water, and then claiming abstraction rights to the water thereby 'saved'.

Chew says that the Phelps Dodge Corporation, which wanted water from the Gila River to expand its Morenci copper mine in Arizona, was among those that employed scientists to talk up the numbers. The US government got involved, since metals from the mines were badly needed to build armaments during the Second World War. Demonizing tamarisk became part of the war effort. In 1944, groundwater hydrologist Thomas Robinson and others at the US Geological Survey prepared a water report 'in an effort to determine how much water could be made available for an essential war industry by removing the salt cedar growth from the bottom lands of Gila River'. It said the alien formed 'dense jungle-like thickets [that] thrived and spread at the expense of nearly all native plant life' and took 75 per cent of the water.[12]

After the war ended, the battle continued to rage against a plant that was branded by one Colorado newspaper as 'a water-gulping, fire-feeding, habitat-ruining, salt-spreading monster.'[13] A military assault began, with flame-throwers and, later, Agent Orange deployed in the assassination of tamarisk. And USGS's Thomas Robinson 'built a career on tamarisk-bashing', says Chew. In 1951, he told the American Geophysical Union that across the seventeen western states of the US, the plant was wasting water equivalent to twice the annual flow of the Colorado River through the Grand Canyon. He failed to mention that the cottonwood and willow trees consumed nearly as much. But no matter, Robinson's story became the received wisdom, and largely remains so today.[14] Joe DiTomaso, a plant ecologist at the University of California, Davis, reckons tamarisk consumes 'almost twice as much as the major cities of southern California'.

Between 2005 and 2009, US Congress authorized $80 million to uproot and poison the dreaded weed. Researching the parlous state of the Rio Grande at that time, I was repeatedly told how a stretch known as the 'forgotten river' downstream of El Paso in Texas had been dried up by tamarisk. Each scrubby indestructible bush sucked up as much as a thousand litres of water a day. The virtue of clearing tamarisk from rivers and their banks – whether by ripping it up or dousing it with herbicides – was one of the few things that river engineers, miners and environmentalists all agreed on. All believed it would deliver them more water.

But would it? What is the truth about tamarisk? Chew, who describes himself as a 'reformed xenophobe', accuses his fellow scientists of peddling outdated and distorted data to 'present the species as an extreme or unnatural agent'. Certainly, it is tough, he says. Certainly, it is a common plant in the arid West. By some estimates it occupies 400,000 hectares, mostly along river courses. It will survive almost anything a desert can throw at it. But that does not make it to blame for the desert being there. Chew says the plant is a scapegoat. The real water guzzlers are the miners, farmers and cities. During my journey along the Rio Grande, I had noted that the river was empty as it left El Paso, long before it reached the tamarisk-infested 'forgotten river' downstream. Perhaps cause and effect were being confused. Perhaps tamarisk is being blamed for desert conditions simply because it can thrive there while other species cannot.

The hydrology now begins to back this up. After an extensive eradication of tamarisk along the Pecos River in Texas over five years from 1999, Charles Hart of Texas A&M University could find no evidence of any greater flow in the river.[15] Early estimates reckoned that with a single plant consuming 760 litres of water a day, a dense stand of tamarisk drew down the water table by about 5 metres a year, but more recent data put its drawdown potential at around 1 metre a year. Tamarisk 'has taken the rap unfairly', says Chew. It is no more a water thief than cottonwood or willow or a host of much-loved natives.

After a reprieve from hydrologists, tamarisk is also getting a makeover from ecologists. While the environmental orthodoxy still holds that tamarisk crowds out native species, the recent evidence suggests that it simply moves in if others disappear. Interestingly, it does best on river courses downstream of large dams that eliminate spring floods. Where the floods on untamed rivers persist, cottonwood and willow continue to dominate. Below dams, tamarisk prevails. On this reading, it emerges not as a menace to other species but as a colonist able to step in where others fail.

Meanwhile, ornithologists have discovered that in many places, tamarisk is the preferred nesting site for a wide range of birds, most notably the endangered southwestern willow flycatcher (*Empidonax traillii extimus*). Tamarisk and the flycatcher share similar territory in the arid West, which probably explains why trigger-happy conservationists were quick to blame it for the bird's decline. But in some areas, 75 per cent of the endangered sub-species nest in tamarisk bushes. Downstream of Glen Canyon dam on the Colorado is a notable instance. Overall, a quarter of the flycatchers nest in tamarisk.[16] With the birds down to fewer than 500 breeding pairs, this is a big deal – especially as conservationists say the main threat to its survival is loss of habitat. Nor can tamarisk be accused of being the only option for birds that would prefer native trees to nest in. Mark Sogge at the USGS found that the flycatchers fed as well, reproduced as well, and survived as well on tamarisk as other trees.[17] Right now tamarisk looks the best habitat for the bird. Conducting a war against it doesn't sound smart. Soon they may be planting tamarisk again.

*

Since the first Europeans showed up, it has always been morning in America for alien species. Like humans, they just keep on coming. In 1672, English traveller John Josselyn catalogued 40 European weeds that, he said, had 'sprung up since the English planted and kept cattle in New England'. Early and generally benign arrivals included buttercups and ox-eye daisies. By the 19th century, New England was known as 'the garden of European weeds'.[18]

Botanists quickly learned where to root around looking for the most interesting newcomers. The best botanical Ellis Islands included the dumps where ships from Europe emptied their rock and soil ballast. Purple loosestrife (*Lythrum salicaria*) probably arrived this way two centuries ago. The pretty flower likes damp places and watersides, and spread far and wide along the developing network of navigation canals in the East and of irrigation systems in the West. Now it can be found on river banks and in wetlands from the Hudson River to Alaska.[19] It is widely seen as a threat to biodiversity, especially in wetlands, though the evidence is thin, says Michael Treberg of the University of British Columbia.[20] And native bees like it, which is a plus in some eyes.

As the American frontier moved west, many European plants followed. They included grasses eager to take up residence in the wide open spaces. Among them was cheatgrass (*Bromus tectorum*). This old-world grass fed some of the first domesticated animals, thousands of years ago in the Middle East. It is now a truly global citizen, found from New Zealand to Greenland and South Africa to Japan. It arrived in America with livestock, making a first appearance in Pennsylvania in the 1860s. 'After that it came west with the stagecoaches, and began taking over from sagebrush', says DiTomaso, author of the standard textbook on the weeds of California. It travelled well. Its hairy seeds blew in the wind, attached to clothes and got hooked into the hairs of cattle and sheep. It also mixed with the hay that fed the animals pulling coaches and wagons.

Cheatgrass likes things out West. The knee-high yellow grass is today the most common plant in the region, covering millions of hectares, especially in Idaho, Nevada and Utah.[21] It steals a march on native rivals by germinating early in spring. That also makes it a valuable fodder crop. The downside is that it dries out by mid-June, and burns well. Its dense growth also means that fires travel much faster than in the days when sagebrush dominated. 'In Idaho they have million-acre burns', says DiTomaso. After the fires, cheatgrass recovers first and takes over more land. Thus cheatgrass encourages fires, and fires encourage cheatgrass.

Other old-world grasses that have done well include couch grass, foxtail and meadowgrass (*Poa pratensis*). A principal reason why all four have prospered is that they were well-matched with the grazing animals that they accompanied westwards. European cattle are much more voracious grazers than the bison they mostly replaced. Native sagebrush could not cope. European cattle required European grasses. Only their dense growth could withstand the assault of a million teeth. Cheatgrass domination is often described as a step towards desertification, but almost certainly it held back the soil erosion and desert spread that would have followed long-term grazing of sagebrush by European cattle. Americans probably have reason to be thankful for the foreign grass.

When it comes to nature, we are fickle. We forget that much that appears loveably native is actually foreign. Some vagabond species that we like can become so much part of the landscape that they get local names and even become adopted natives. In the US, old-world meadowgrass was renamed Kentucky bluegrass. Tumbleweed sounds quintessentially American – a metaphor of ghost towns in Western movies as it rolls across the landscape in the wind. But the most common tumbleweed species is not native at all. *Salsola tragus*, Russian thistle, was brought by Ukrainian immigrants in flax seed.

Likewise, few would regard the humble and helpful earthworm – 'nature's little farmer', ploughing and fertilizing soils – as an alien from the old world. But it is. Canada and the northern states of the US had no earthworms until Europeans showed up. There are several species around now, including the most numerous, the litter-dwelling octagonal-tailed *Dendrobaena octaedra*, and Europe's favourite, *Lumbricus terrestris*. Alien they may be, but they are perhaps only replacing worms wiped out before the last glaciations scraped away the soils of the north – the final piece in the reconstruction of post-glacial soil ecology.

Not everyone accepts the rise of European worms. Some US forest ecologists say that before their takeover of American soils, forest floors were thick with leaf litter. Now that worms eat this litter, soils are exposed and species have disappeared, including some salamanders and ground-nesting birds.[22] One lifelong worm enthusiast, Paul Hendrix of the University of Georgia, thinks it is worth trying to hold back their advance.[23] Time, at least, is on his side. Typically, invasion fronts advance by a kilometre a century, says Cindy Hale of the University of Minnesota, a fanatic who started a worm farm as a kid.[24] So, despite many introductions, their conquest is far from complete. Outside urban areas, they are mostly found near roads and lakes – suggesting introductions on vehicle tyres and as fishing bait. In Alberta, Canada, 90 per cent of boreal forests remain worm-free.

Another unexpected alien is the European honeybee (*Apis mellifera*). The species had been domesticated in the old world for four millennia before British settlers brought fully stocked hives in the 1600s. Some ecologists see the European arrivals as an ecological disaster. But on a continent where the main source of sweetener had previously been maple syrup, honey from the aliens was welcome. And while there were pollinators in North America before, the European bees have assumed a major role, especially for crops. Today, they do an estimated 80 per cent of the pollination carried out by insects on US farms and in orchards – and a third of all pollination.

Most Americans today see the European honeybee as fully naturalized. Twelve states list it as their state insect. When nearly a third of honeybee colonies died in 2012, from a mysterious disorder, there were no cheers for the demise of an alien invader. Rather, the media pointed out the honeybee's $20 billion-a-year value to the US economy. (And, curiously, at this point it became known as an American bee rather than a European.)

The African killer bee is not so welcome, and is likely to be long regarded as an alien. It is a hybrid that was created almost 60 years ago when a Brazilian beekeeper in São Paulo accidentally released a couple of dozen Tanzanian queen honeybees (*Apis mellifera scutellata)* from his hive. They mated with European honeybees (*Apis mellifera)* and the resulting Africanized European honeybee has been moving north ever since. The name 'killer bee' arises not because they kill humans, but because they invade the hives of European honeybees, killing the queen and installing their own matriarch in her place. After reaching the US in 1990, they have spread through the South. They are probably unstoppable now, but sadly they don't even produce much honey.

Invading pests and diseases are a special problem for the US. Tree pests in particular find that American forests often provide easy pickings. The gypsy moth was brought to the US in 1869 by Étienne Léopold Trouvelot, a French republican and amateur entomologist. He wanted to cross it with silk worms to create a profitable textile. Instead, some larvae escaped into the woods behind his home in Medford, outside Boston. They have been eating their way through the hardwood trees of North America ever since.

The boll weevil moved north from Mexico to ravage the US cotton crop in the 1890s. Around the same time chestnut blight, a fungus, hitched a ride with some Asian chestnuts brought in for nursery stock by one of the US government's early botanical explorers, Frank Meyer, who is most famous for a Chinese lemon introduced to the US and named after him. Chestnut blight has since virtually eradicated indigenous chestnuts. In recent times, a green beetle called the emerald ash borer showed up in packing material on a ship from Asia that docked near Detroit. It has since been laying larvae in the bark of ash trees. Once they hatch, the larvae bore into the tree and kill it.[25]

Nobody would make a case for wanting to embrace such troublesome arrivals – though stopping them at the border is fiendishly hard. But amazingly to some, there may be an argument for giving at least a tepid welcome to another *bête noire* of American conservation, the zebra mussel (*Dreissena polymorpha*).

As a general rule, aliens take hold first and best in disturbed places, whether the result of hurricanes or floods or human habitation. Few places in the mid-20th century were as disturbed by human presence as Lake Erie, the smallest and most industrialized of the Great Lakes. Erie's ecology – or rather, its absence

– was headline news. The lake had been declared 'dead' after a heavily polluted river entering it, the Cuyahoga, caught fire several times in the 1960s. It was never quite dead, but pollution had killed off most of Erie's once-remarkable collection of 300 native species of mussels.

Then, in the mid-1980s, a foreign mussel showed up. The zebra mussel was stuck to the hull of a ship from the Caspian Sea. It was arguably a just, if accidental, retribution for the US export to the Soviet empire of *Mnemiopsis*, the 'blob that ate the Black Sea', as we saw in Chapter 3. The ship came up the St Lawrence Seaway, and the mussels loosed their grip in the tiny Lake St Clair. By the summer of 1989, they had slipped next door and were all over Lake Erie, showing what Soviet mussels could achieve in an environment that had wiped out the natives.[26]

They don't look obviously tough. Zebra mussels – so named because of their distinctive stripes – are tiny and largely immobile. But they spread easily, attach to any surface and can attain extraordinary concentrations. This invasion was a very collective endeavour from the land of collectivism. Most importantly, they didn't mind the lake's pollution. As filter feeders, they ran vast amounts of the lake's vile water through their bodies in order to extract the plankton, their main food. This was good news for the lake. Zebra mussels turned out to be the best janitors Erie ever had. They were almost the only filter feeders able to swallow the gunk and survive. Each of the tiny animals was gulping down, filtering and excreting as much as a litre of water a day. Billions of the creatures can filter a lot of water that way. During the 1990s there were probably enough of them to process the lake's entire volume in a week. Along with the plankton, they ate up most of the pollutants suspended in the water, excreting it onto the lake floor.

Native mussels, clams and the shrimp-like *Diporeia* have declined because the plankton is being consumed by the zebra mussels. And there are fouling problems on the lake floor from the excretions. But it has to be said that since zebra mussels became established, the clarity of the once-murky water in Lake Erie has increased dramatically. You can see down for 10 metres in some areas, compared to less than 15 centimetres half a century ago. As light has penetrated the lake, some aquatic plants have revived. They in turn have become nurseries for fish such as the yellow perch.

The zebra mussel is also itself a food source for important species. The smallmouth bass (*Micropterus dolomieu*) and, most dramatically, the previously endangered lake sturgeon (*Acipenser fulvescens*) – a giant shark-like beast that has barely evolved for 100 million years – munch them, and have revived their populations as a result. Lake Erie is now reportedly the world's premier smallmouth bass fishery. Meanwhile, migrating ducks that once avoided the fetid waters now make detours to feast on the new mussels.[27]

If you are of a gloomy persuasion, you might sympathize with the complaint of invasion biologist Daniel Simberloff that all this clearing of Erie waters 'favoured invasion by large-leaved aquatic plants such as Eurasian water milfoil'.[28] That seems perverse. A new ecological environment is rapidly emerging in the lake, with its own checks and balances. We cannot tell quite how it will work out. The revival of the ancient sturgeon may limit and perhaps even reverse the proliferation of zebra mussels. Some fish ecologists say that if humans hadn't hunted the sturgeon almost to extinction the zebra mussels would never have got such a grip. So the battle for supremacy between the two species might now get interesting.[29] But surely all this is better than the 'dead' lake.

None of this denies that the mussels can be a physical menace for human activity. They cling onto any hard surface they can find, congregating in dense masses many layers thick. They cover jetties and clog water intake pipes. The extent of the problem emerged when Lake Erie's main port town, Monroe, lost water for three days. But we should be wary of some cost estimates floating around. One widely quoted statistic puts the cost to utilities and others in the Great Lakes alone at half a billion dollars a year. It is often referenced to the Center for Invasive Species Research at the University of California, Riverside. But when I checked it out, the centre's director Mark Hoddle said they had been quoting someone else, but he couldn't tell me where they got it from. Maybe it would be better to go with the figure from Nancy Connelly at Cornell. She found, after interviewing corporations, that the zebra had cost just a quarter of a billion dollars over fifteen years. That is still substantial, but only one-thirtieth as much as Hoddle's annual figure.[30]

What happens in Lake Erie does not necessarily stay in Lake Erie. Success breeds success. In less than a decade, zebra mussels have spread to all five Great Lakes and on into connecting rivers including, via Chicago's canal system, the giant Mississippi system. They are now in 27 US states. In places this has been good news. 'Zebra mussels have had positive impacts on parts of the Great Lakes ecosystems', says the USGS's Great Lakes Science Center.[31]

As it bobs down the Mississippi, the zebra mussel will probably pass one of the various species of Asian carp swimming the other way. Asian carp (*Catla catla* and others) can grow over a metre long and eat 40 per cent of their own weight in plankton daily. Imported from China in the 1970s to eat algae in catfish ponds and sewage works, they escaped and spread up the Mississippi. There are fears that this infidel may soon find its way along the canals linking the river through Chicago to Lake Michigan and the other Great Lakes.

All manner of ruses are being tested to keep it out, including chemical lures, a high-voltage electric barrier, an acoustic gun that acts like an aquatic bird scarer, and simply blocking the canals. The Army Corps of Engineers is advising the White House. A more enterprising option is to treat the carp as a resource and get eating. One popular Chicago fish bar, Dirk Fucik's Fish and

Gourmet Shop, has put carp burgers on the menu. Meanwhile, despite the angst, nobody knows if the carp would be a threat if it reached the Great Lakes. Some were discovered swimming in Lake Erie a decade ago, but disappeared again, suggesting that an ecological takeover is unlikely.[32]

Somewhere too, the striped mussel will surely meet the northern snakehead (*Channa argus*), a fish with 'teeth like a shark' and a voracious appetite, which can wriggle overland like a snake for up to three days. This Chinese 'Frankenfish' first grabbed America's attention in 2002 when it turned up in Maryland. The story is that an Asian-born resident of Maryland ordered a couple of live snakeheads from a Chinatown store in New York to make a traditional medicinal soup for his sick sister. But by the time the fish arrived, his sister was well again. So he threw them into a nearby pond. There have since been sightings across the country. But the idea of a wriggling scourge spreading across the US is probably far-fetched. Most sightings – including those in Lake Michigan and California – appear to have been the result of individual local releases. But we naturally prefer the narrative of US interior secretary Gale Norton as he banned the movement of snakeheads across state lines, that their odyssey was turning into a 'bad horror movie'.[33]

*

Following the snakehead, I went to California, which has more than a thousand species from out of state. But only 30 native species are known to have become extinct as a result. The resulting ecosystems are a cosmopolitan mishmash, which means any full-scale assault on aliens would create a bloodbath.

Susan Schwartz, in her laid-back Californian way, knows this. She is president of the Friends of Five Creeks, whose self-imposed task is eradicating aliens. The creeks of which she has informal charge are all around Berkeley and drain into San Francisco Bay on its east side. All contain alien species in their water or along their banks. The Friends brand themselves weed warriors. They go out most weeks and rip up invaders. When I visited, Schwartz was on the hunt for Algerian sea lavender. 'It arrived in 2006 and is exploding', she said. It has 'some very pretty flowers', but it competes with native salt marsh species.

Her other bugbear is French broom (*Teline monspessulana*), which is both invasive and a fire hazard. The tall evergreen with yellow flowers took out over 100 metres of waterfront in Berkeley. Ripping it out is satisfying, she says, but the seeds stick around and the Friends have to go back every year to remove new seedlings. Also on her hit list are southern Europe's pepperweed, German ivy, Japanese dodder, Italian ryegrass, the Hottentot fig from South Africa, Pampas grass from Argentina, English ivy and Himalayan blackberry (*Rubus discolour*). But on second thoughts, she rather likes that last one. It's not much of a problem. It shows up on waste ground and people go out and pick the berries. It's useful. Let it be. But the English ivy has to go, more because it hides broken glass and dog poop than for any ecological imperative, she says. I liked Susan. Her pragmatism was almost European.

By a shopping mall we check out Cerrito creek. It isn't a natural creek at all. 'They filled in a marsh here for the mall and car park, and put in a creek to drain it', says Susan. Blackberry bushes soon turned up, clogging the channel and providing cover for the local kids with their BMX bikes and pot smoking. The bushes have been removed – but more, I suspect, as part of a social cleaning of the neighbourhood. Local officials are putting up signs so you can recognize the creek's bird life, and regular citizens turn up now for jogging and tai-chi. This is California, after all.

The weed warriors are doing neighbourhood gardening rather than the eradication of alien species. Some aliens must stay. A third of California's native butterflies depend on non-native plants for their food.[34] The monarch butterflies would be in trouble without the state's alien eucalyptus trees, and the bees need eucalyptus nectar during the winter. 'We know we can never get back to nature as it was', Susan says. 'There is no balance of nature, no climax ecosystem that nature wants to revert to. But we just want to stop takeovers. That's all.'

San Francisco Bay is the most invaded estuary in the world. Or so said Andrew Cohen, then of the San Francisco Bay Institute, and James Carlton of Williams College in Mystic, Connecticut in *Science* in 1998, and nobody has contested it.[35] 'Exotic species dominate many of the ecosystems', they said. The bay contains Amur clams from Russia, Atlantic green crabs, Black Sea jellyfish, Chinese mitten crabs, Japanese gobies and a New Zealand isopod called *Sphaeroma quoyanum* that burrows into mud banks. The eastern mudsnail has replaced some native snails, and carries a flatworm that nibbles at the skin between your toes, causing 'swimmers' itch'. Mediterranean mussels cling onto rocks. There are foreign clams, barnacles, seasquirts, worms, sponges, hydroids and sea anemones.

Since the Gold Rush in the late 1840s and 1850s, there has been a biological rush into the bay, and foreign sea life continues to pour in beneath the Golden Gate – stuck to ships' hulls and oil rigs, carried in ballast tanks and hidden in cargo unloaded at the ports of San Francisco, Oakland and Sacramento. More than half the invasions have happened since 1960, with new arrivals since then averaging one every fourteen weeks. 'Drilling platforms and oil rigs can bring whole communities of organisms', Cohen told me when we met at his house in the hills overlooking the bay. And because the rigs stay, the whole community may hop off and take up residence. Other sources of accidental releases include discharges down rivers and drainage pipes. The Japanese anemone may have come from local aquariums. The international trade in fish bait has brought marine worms from Europe, Japan and even Vietnam. Deliberate releases include striped bass, which arrived in rail cars from the east coast in 1870 to establish a local game fishery, and Atlantic and Japanese oysters, which have been farmed in the bay since the late 19th century. With the oysters came a range of parasites and pathogens that are still around.

Most fish and shore birds in the bay are still natives, says Cohen. But the food they eat has changed dramatically. There are at least 300 alien species established in the bay. Some invasions happen very fast. Cohen remembers that 'in the fall of 1986, a college field class found three new clams in a sample of mud on the bottom of the bay soon after a big spring flood'. One of them, a Chinese clam called *Potamocorbula amurensis*, had within two years become the most abundant clam in the southern bay, carpeting large areas of mud and sand and making up 95 per cent of the biomass.[36] 'It would have been filtering the entire volume of the South Bay every day', says Cohen. It was eating all the plankton, so the plankton blooms typical of the bay simply stopped. Meanwhile, native copopods (crustaceans that drift in the water) disappeared from parts of the bay, and were replaced by Asian versions. Fisheries also collapsed.[37]

But was the arrival of the Chinese clam to blame? It may sound a clear-cut case, but Cohen is not sure. The 1986 spring flood that preceded the Chinese invasion dramatically freshened the bay water. That wiped out most of the dominant Atlantic clams, which were themselves 19th-century interlopers. Cohen says that the Chinese clam was probably lurking in the bay all along and simply took advantage.

The key point about the bay is that it has never been a stable ecosystem with a fixed set of species. It is subject to occasional dramatic spring floods that course down the rivers and flush the bay with fresh water. When that happens, says Cohen, the entire ecosystem is rebooted. Jay Stachowicz, a marine biologist at UC Davis, agrees: 'The bay is dynamic, constantly being purged by freshwater floods. It is after those events that introduced species make their move, but if they didn't something else would.'

In fact it is this dynamism that makes the bay what it is. Those conservationists who bemoan every change in the bay, every arriving alien species and every disappearance, are rather missing the point. The constant change is the reason why, despite more than a century of bombardment by alien species, San Francisco Bay is in such rude health. Don't take my word for it. In early 2013, the international Ramsar Convention on wetlands accepted the US government's designation of the bay as of 'international importance'. It called the bay 'a natural wonder and a critical ecological resource', one of North America's most ecologically important estuaries. Yes, 300 of the thousand or more species of fish, mammals, birds, plants and invertebrates in and around the bay are recent arrivals. But that is something to celebrate, a sign of health rather than sickness.

The bay may well be, as Cohen and Carlton claimed almost two decades ago, the most invaded estuary in the world. But that, in reality, is the key to its other defining feature, what the Ramsar secretariat agreed was its 'superlative biodiversity'.[38] 'Change is the norm here', says Cohen. 'It's the story of California, for species as well as humans', says Stachowicz.

*

Another dynamic American ecosystem is the Florida Everglades. It too has been overrun by alien species in the past century, rather like San Francisco Bay. Today, a quarter of all the fish, reptiles, birds and mammals in the Everglades are exotics. But it is hard to work out what we should regard as native, or why anyone should imagine the area would show any kind of natural balance and ecological stability. The wetland is recent and has experienced constant changes in sea level. Even if humans had stayed away, it would still be in a perpetual state of flux, and a refuge for migrant species.

But I agree that is not an easy message to digest when the big-daddy newcomer in the Everglades is something as big, aggressive, hungry and foreign as the Burmese python (*Python molurus bivittatus*), one of the world's five largest snakes. There are an estimated 30,000 of them lurking in its wetlands. The largest of them are able to catch and eat the wetland's former top predators, alligators.

How did they get there? Florida is a long way from Burma. An obvious explanation is the craze for pet snakes in the 1990s, when numbers in the Everglades took off. Maybe, as they began to reach their full size of up to 5 metres, scared owners got rid of them – down the drain, into the nearest creek, or after a car-ride to the swamp. But even though the Everglades are a nice warm and wet habitat – not so different from back home in the marshes of the Irrawaddy – sporadic pet releases seem unlikely to be to blame for such a large population. The most likely source is a mass escape that happened when Hurricane Andrew ripped through southern Florida in August 1992. It wrecked a warehouse in Homestead, a Miami suburb right on the edge of the Everglades, where they bred pythons for pet shops. The walls fell in and tanks were smashed. Reportedly, 900 pythons of varying sizes were living there. The baby ones may have been literally blown into the Everglades. Big ones perhaps just slithered away into the swamp.[39]

Anyhow, they got there. And the giant constrictors eat a lot. Besides deer and the odd alligator, they hunt down and gobble up raccoons, rabbits and opossums. Once the Everglades' most frequently seen species, all three are vanishing fast. A survey of road kill made by park rangers found a decrease of 90 per cent and more in their numbers over the past twenty years.[40] These animals may know enough to dodge alligators, but they have no previous experience of large constrictor snakes. Birds also make a big part of what vets find in the stomachs of captured pythons.[41]

Meanwhile, a half-hearted and probably doomed hunt is on to eliminate the monsters. This largely comprises offering prizes to hunters. No doubt the participants have fun, cruising through the swamps in search of big snakes to shoot. But over ten years, fewer than 1,800 have been caught. And the Python Challenge in early 2013, which was entered by 1,600 snake-wranglers, bagged just 68 reptiles. That seems a poor return for a lot of effort. Probably we had best accept that they are there for good – or until nature thinks differently. As food stocks subside, we should expect python numbers to stabilize and start falling. They may settle down to life as one of a range of Everglades snakes that includes the eastern diamondback, the world's largest rattlesnake and the most dangerous snake in America.

An estimated quarter of all the animals in the Everglades are exotics. Caribbean and Central American visitors include green iguanas from Puerto Rico, Cuban tree frogs, Jamaican fruit bats and Mexican squirrels, together with the odd armadillo from South America. From Africa, there are a bunch of Gambian pouched rats that escaped from a local keeper on Grassy Key, and around a thousand Nile monitors. Most of these 1.5 metre-long flesh-eaters are probably also expelled pets.

And then there are the plants. Most of the Everglades used to be treeless. Now the famous rivers of grass have in many places been replaced by Brazilian pepper-trees and Australian paperbark trees. Both were introduced a century ago to dry out the swamp and now together cover some 200,000 hectares. These are major invasions, with big impacts. But with the Everglades always in flux, ever dependent on sea levels, species have constantly come and gone. There wasn't a Florida Everglades at all until about 5,000 years ago. It has never settled down to any fixed form, going from mangrove to cypress swamp and hardwood forest and on to the familiar seas of sawgrass, as sea levels rose.[42] Species came and went. They still do.

*

The bottom line here is that we should treat species on their merits and learn a little tolerance and respect for foreigners. They have their idiosyncratic and occasionally disruptive ways, but they add to the variety of life and often bring benefits. And they certainly aren't going away. There are an estimated 5,000 alien plant species in US ecosystems, almost a third of the total species count. Florida has 900 of them, and California 3,000. There have been extinctions among the natives, but remarkably few. The aliens bring biodiversity, a fact that is rather inconvenient for the many conservationists who argue vehemently that invaders damage biodiversity. So, in what seems to me an extraordinarily Orwellian piece of science, they have been trying to define away the issue.

Here is how they did it. In 1997, eleven leading American conservation scientists headed by Edward O. Wilson, Daniel Simberloff, Peter Vitousek and Stanford's Harold Mooney, wrote to US vice president Al Gore arguing that 'a rapidly spreading invasion of exotic plants and animals ... is destroying our nation's biological diversity'. This is not true. But no matter. They argued that to end this non-existent loss of diversity, there should be a 'national program to combat invasions'.[43]

They got their wish two years later. An executive order created the National Invasive Species Council, charged with ending the 'environmental harm' done by aliens. Heavily advised by the aforementioned scientists, the Council defined harm as 'biologically significant decreases in *native* species populations, alterations to plant and animal communities or to ecological processes that *native* species and other *desirable* plants and animals and humans depend on for survival'. (My italics.)

This was a problematic definition. It meant that almost any change – any 'alteration to plant and animal communities' – was regarded as harmful. A great many ecologists disagree profoundly with that. Change and 'alterations' are natural and the norm. This definition also means that any damage to native species is found to be harmful, whether or not it has any impact on overall biodiversity. In fact we seem to have gone a long way from any interest in biodiversity. The interest is entirely to do with protecting natives and avoiding change.[44]

This seems to me a perversion of ecology, a stitch-up against alien species. It was compounded the following year when many of the same top conservationists, headed by Harold Mooney, published a statement on the future of global biodiversity for the 21st century in the journal *Science*.[45] It amounted to a clarion call to protect biodiversity. Who could quarrel with that? Except that, in the first paragraph, they declared that 'our definition [of biodiversity] excludes exotic organisms that have been introduced'. They did not explain or justify this statement. I find it extraordinary to arbitrarily exclude one large and growing element of biodiversity.

For them, alien species don't count and are not counted. They do not exist as part of nature. They have no place. They are un-nature, if not anti-nature. They should be gone. Under this definition, biodiversity in the 21st century can only go down. Extinction could cut the number of species, but introductions could never increase it. Thus the inconvenient fact that alien species actually increase real biodiversity in many places is simply defined away. Big Brother in *Nineteen Eighty-four* would be proud. Franz Kafka would be proud. Joseph Heller would have added an ecological chapter to *Catch-22*, if he had known. It sounded more like an ideology than good science.

5

Britain: A Nation Tied in Knotweed

Just as Britons took their favourite species with them to their imperial colonies, so they have always brought exotics home. There are some 2,300 alien species in Britain. It is now reluctant host to American mink that escaped from fur farms; ruddy ducks from Canada that took flight from conservationist Peter Scott's Slimbridge wetland; Chinese mitten crabs in the Thames, so called for their furry claws; and North American signal crayfish that swam free from aquaculture ponds and exterminated native crayfish. But, as a nation of gardeners, Britons save their special wrath for an ornamental shrub from the slopes of a Japanese volcano.

The South Wales city of Swansea is sometimes called the Japanese knotweed capital of the world. The Japanese might disagree, but the plant (*Fallopia japonica*) has certainly made itself at home here. In summer its jungle of bright green heart-shaped leaves and pretty white flowers covers valley sides, chapel cemeteries, and even the sand dunes down on the beach. In winter its bare stems give the whole city a reddish tinge. Sean Hathaway, who works for Swansea city council, is Britain's – possibly the world's – only full-time Japanese knotweed officer. Hathaway welcomed me at the station as one more of a steady stream of journalists from around the world who come to report the horror story. Investigative titans at the *Sunday Times*, *Harper's* essayists and bemused Japanese scribes all come to South Wales to walk round the 'JK jungles'.

'We think the weed came to Swansea as long ago as 1918. We only know because we found a picture postcard of a church with a bush in the foreground. But nobody noticed it until the 1970s, when people began to complain about this weed invading their gardens.' There was a reason for the weed's sudden virulence. The city was for a long time the centre of the world's copper industry. But as the works shut, it became instead one of the largest areas of industrial dereliction. Knotweed likes dereliction and has no problem with metal pollution. 'It can survive in virtually any soil type', said Hathaway. It once flourished on lava fields, after all. It colonized land that other plants didn't fancy, and once it got going it tended to crowd others out. The dereliction was followed by one of the biggest urban renewal schemes in Europe. The massive volumes of earth being moved spread the knotweed rhizomes, which have been consolidating ever since.

Japanese knotweed was a novelty even in Japan. It specialized in sending down deep rhizomes to find nutrients beneath lava flows. It came to Europe with a Bavarian doctor called Philipp Franz Balthasar von Siebold. Assigned to Japan by the Dutch East India Company in the 1820s, he indulged his love of plants, often taking them as payment for surgical procedures. He shipped his favourites home, where they were planted out in his employer's botanical gardens in Leiden.

Samples reached Britain's equivalent, Kew Gardens, sometime in the 1850s, and were widely sold to commercial nurseries.[1] Rather like kudzu in the US, its Japanese origins made it an attractive novelty. The celebrity gardener of the day, William Robinson, was a fan. An Irishman, Robinson was famous for rebelling against the military monocultures of flower beds of the time. He wanted a cacophony of bright, expansive plants growing together in a less tidy and more natural way. His 'wild gardens' teemed with exotics gathered from across the globe. He brought goldenrod and Michaelmas daisies from North America, for instance. He liked lilies and pampas grass and Japanese anemones. In his most famous book, *The English Flower Garden*, which was first published in 1870 and constantly updated with his new finds, Robinson called Japanese knotweed 'a plant of sterling merit ... undoubtedly one of the finest herbaceous plants in cultivation'.[2]

Several of Robinson's 'wild plants' showed their wildness by growing rather too well. When they proved too much and were thrown away, there were predictable results. A plant like Japanese knotweed could thrive almost anywhere it found space and sunlight. It took to waste dumps, railway embankments and graveyards across the land. It was growing wild in London by 1900 and across Yorkshire by the 1940s. By the 1960s, it had settled onto the barren moorlands of the Scottish Hebrides.[3] As it travelled, its reputation changed from Honshu darling to scourge of the land. Today, most gardeners would agree with the Royal Horticultural Society that it is 'a pernicious pest'. Local TV news stations revel in stories of how it bursts through the foundations of houses, poking into living rooms and spreading like some malign houseplant. Reports often include interviews with scientists wearing lab coats and masks to explain how nasty it is.[4] But how much of a scourge is it? Does it pose an economic threat? And how widely has it spread in the countryside?

The British government's Environment Agency calls JK 'indisputably the UK's most aggressive, destructive and invasive plant'.[5] It 'shades native plants and damages hard structures such as tarmac, paved areas and flood defences', and is classified alongside the most toxic chemicals as a 'controlled waste'. Landfilling Japanese knotweed is a crime.[6] The government says the weed costs the British economy a staggering £170 million a year – or a pint of beer for every person in the land. But I can find no evidence that anyone spends even a small fraction of that amount on fighting Japanese knotweed. The Environment Agency, the country's largest land manager, spends only £2 million a year killing and uprooting it. Swansea council spends a paltry £30,000 a year. It turns out the costing is an estimate from CABI, a former government agency that is now a non-profit consultancy and invasive species eradicator. Dick Shaw, its deputy director – and often the lab-coated scientist in those TV news scare stories – told me it was an extrapolation from data collected in Swansea.[7]

The calculation goes like this. Swansea, as a recognized hotspot for the weed, is the only place in Britain to require building developers to certify there is no knotweed on land before they get planning permission for building. So if they have any JK, they have to exterminate it. Swansea says that just under 3 per cent of its planning applications require treatment. The typical treatment cost is £5,800. There are on average 568,000 planning applications per year for building across Britain. But here is the bit I find impossible to accept. CABI decided to assume that the national rate of infestation would be similar to that in Swansea, at 2.25 per cent. On that basis it reckoned that nationwide 12,800 applications should theoretically need treatment. It then doubled the treatment cost to allow for legal costs and other extras and multiplied up to produce a national cost of Japanese knotweed of just under £150 million a year. Or £165 million, to allow for expenditure by public bodies on river banks, roadsides and railway embankments and potentially lowered property values.

I find such a high figure extremely difficult to justify. Swansea is, almost everyone agrees, exceptional. Few other authorities impose planning requirements for knotweed, because it is not a problem to them. CABI offers two reasons to justify assuming that the rest of the country is like Swansea. The first is that 'much of the country will be behind Swansea in Japanese knotweed management'. In other words, other places may not be spending that kind of money, but they should – and probably will one day. The second is that, as Shaw put it to me, Swansea 'was the only place in the country that had any figures at all that could be scrutinized'. As fuzzy as this line of reasoning is, it is what CABI's figures are based on, and other professional bodies I spoke to seem willing to accept it.

It also justifies Shaw's work developing a 'biocontrol' for the weed. From his CABI headquarters down a leafy and JK-free lane off the A30 in the Surrey commuter belt south of London, he has been working on deploying a louse called *Aphalara itadori* that he collected on Japanese mountains and nurtured in his labs. But after three years of trials at eight sites, including one near Swansea, he had to report that 'each time we have had a weather disaster: cold nights, late springs or wet summers'. The louse always died. He was thinking of going back to Japan to get a new stock of lice, or of pushing the government to allow bigger field trials.

My scepticism about CABI's calculations does not mean there is no issue. There is. The Swansea locals hate Japanese knotweed. It can blight their lives. But not because of any damage it does to their homes or gardens. That is minimal. The problem, everyone in Swansea I spoke to said, was the panic it creates among banks and others who might finance loans to buy houses. If there is a hint of knotweed, some will not lend at all. Japanese knotweed was a problem weed that needed careful handling, but no more, until five years ago when media stories started to appear. Then it became 'the scourge that could sink your house sale', according to the *Observer*. 'Britain is in the grip of an alien invasion', said the *Daily Mail*. A 'race of female clones ... with giant root systems that burrow down nine feet and towering bamboo-like stems that grow four inches a day' was spreading across the land.[8]

'Mortgage lenders started to go spare. Some won't lend at all now if there is knotweed on the property', said Hathaway. 'It's ridiculous. The houses are as valuable as they ever were.' At the very worst, if the weed was pushing its way into the living room, it would cost a few thousand pounds to get rid of it. That is tiny compared to the value of the house.

Many independent observers take a similar view. John Bailey at Leicester University, who has studied the weed for 30 years, says 'the hysteria about Japanese knotweed is absurd, and is distressing lots of people'. The weed will, he agrees, grow fast after being moved in soil, as happened in Swansea 40 years ago and also more recently during the Olympic construction activities in east London. But it does not spread like wildfire. It will invade cracks, but does not push through solid concrete. It is sometimes a problem weed, for sure. But it is not a £170 million-a-year problem.[9]

Nor is it an environmental threat. I asked Hathaway if JK had spread into the countryside around Swansea, which includes the scenic Gower peninsula. He laughed at the idea. 'Not at all', he says. The entire infestation in the city and surrounding area is 100 hectares, mostly on public wasteland. The biggest single patch I saw covered 2 hectares on a hillside overlooking the railway. James Dickson, a botanist at Glasgow University, says JK usually stays in urban environments, where the disturbed soils it likes are mostly found. In cities, much wildlife rather likes it. Otters, for instance, are returning to cleaned-up rivers across the country. In Sheffield, their favourite place to hide out is beneath the protective foliage of dense stands of Japanese knotweed along the bank of the River Don. Knotweed has other uses, too. It is a valuable source of nectar for insects, including honeybees. The Japanese revere its medicinal properties as a source of the polyphenol resveratrol. Richard Mabey, the naturalist and national champion of weeds, says the young shoots make a nice vegetable. In the US, Japanese migrants eat it cooked, like spinach.[10] If the rest of us learned to stop fearing it, maybe we could do the same.

*

Horticulturalists share a list of the same 'big four' foreign plant invaders in Britain: Japanese knotweed, Himalayan balsam, rhododendron and giant hogweed. All four got their start with Robinson and the other Victorian wild gardeners. All four do best in disturbed land around people, so they are frequent in suburbia and along footpaths and river banks. That makes them highly visible to casual observers, but they are not widespread in the countryside at large. Yet in different ways, all four are victims of the kind of hostile treatment that no native species would receive. Like crossed lovers, gardeners seem to relish hating what they once loved.

Giant hogweed (*Heracleum mantegazzianum*) first came to Britain from the Caucasus in the early 19th century. It looks like giant cow parsley, and Victorian Britons loved its height. In Norway, they still do. There it is known affectionately as the Tromso palm, after the Arctic town where it grows widely. But Britons have learned to fear the hogweed. Legend has it that it strayed from respectable gardens into the wild a century ago by escaping from Buckingham Palace into London's parks. From there it reached the canal system and its floating seeds spread to the rest of the country.

It has colonized river banks like the Wye and Usk in Wales. Even so, nobody thought much about it until the 1970s, when reports emerged of children getting blisters after touching it while playing. Doctors discovered that the sap from its stem sensitizes skin to sunlight, causing it to redden alarmingly. Cases are rare but can be serious. Hogweed juice could, theoretically at least, cause blindness. The media thenceforth called it a triffid, after the botanical menace in John Wyndham's 1950s science fiction classic *The Day of the Triffids*, which took over the world after humans were blinded. In 1981, giant hogweed joined Japanese knotweed as an officially 'noxious weed'. It is an offence both to plant it and to throw it away.[11]

Naturalists such as Dickson say the ecological case against giant hogweed is again overblown. It may stand tall, but it does not, as often claimed, obliterate any other plants in its midst. In Glasgow, there are famous 'hogweed forests' on the banks of the River Clyde near Kelvinbridge, he says. 'But it's not true that everything else vanishes. You get spring floras, a bit like a woodland, with smaller plants such as lesser celandine growing beneath the hogweed.' Mabey says: 'My own impression is that they're not causing much harm.'[12] No matter; CABI's Dick Shaw is working on a rust fungus to deploy against the menace.

Joining hogweed in the doghouse are rhododendrons. London gardener Conrad Loddiges brought the first samples of *Rhododendron ponticum* to Britain, probably from Spain, in 1764. Later, Kew's legendary plant explorer Joseph Hooker added a host of new species after a trip to Sikkim in the Indian Himalayas. In 1853, he had enough to plant a grotto called Rhododendon Dell at Kew that still attracts visitors today.[13] The various species soon showed a penchant for interbreeding to create hybrids. They were an instant hit with gardeners. The plants love Cornwall's china clay pits. The Eden Project, an ecological theme park constructed in one of the former pits, sells what it proudly calls the Trewithen Hybrid.[14]

While still liked in gardens, rhododendrons are feared in the wild because, though not widespread, they sometimes crowd out other plants. But rhododendrons 'can be surprisingly good for many wildlife species including winter roosts for birds, breeding nightingales and cover for deer, badgers and otters', says Ian Rotherham, an urban ecologist at Sheffield Hallam University.[15] Their dense leaf canopy protects native wood mice from predator birds. Mice numbers increase where the alien evergreen lives, and there is also a measurable rise in the number of weasels lurking in the hope of finding those mice unable to take refuge.[16]

Himalayan balsam (*Impatiens glandulifera*), the last of the 'big four' botanic villains, fits a similar template. Introduced by the wild gardeners for what advertisements of the time called its 'Herculean proportions' and 'splendid invasiveness', Himalayan balsam spread along rivers and waterways. But how far?

In late 2014, the Environment Agency claimed it covered 13 per cent of the river banks of England and Wales, and CABI trumpeted the stat while releasing a rust fungus to wipe it out. But that 13 per cent figure appears rather inflated. When I checked out the source, an old river habitat survey conducted by the Agency, it turned out that to qualify for covering a river bank, it was only necessary for surveyors to find 'a single occurrence' of the plant along a 500-metre stretch of sampled river bank. That is hardly 'covering'. Moreover, the Agency had separately tallied where the weed occurred along more than a third of the bank stretch. The figure for that – surely a much more sensible measure of how much the bank was occupied by the weed – was just 3 per cent.

But while the Environment Agency and CABI clearly have it in for Himalayan balsam, it also has friends. The weed produces excellent pollen that attracts both honeybees and bumblebees. The British Beekeepers' Association says it 'can be crucial in helping honeybees provision their colonies to survive overwinter'. The association advises its members to keep a few bushes in their gardens.[17]

As we have seen, such alien plants like disturbed land, and so congregate in urban areas and other places where we notice them. But they rarely invade the countryside. On a ranking list of common species in the British countryside, the much hated rhododendron came 233rd, with Himalayan balsam, Japanese knotweed and giant hogweed lagging even further behind.[18] In terms of their potential to invade, some natives are a much greater nuisance. Bracken, brambles, stinging nettles and ivy are all natives, and love disturbed ground and nutrient-rich soils often created by human occupancy. Other native species – including holly, gorse, hawthorn, blackthorn and birch – invade grassland, heaths, moors and bogs. If they were foreign there would be a hue and cry. Imagine if the stinging nettle was an alien.

Rob Marrs of the University of Liverpool looked at the prominence in the National Woodland Survey of three local 'thugs' – bramble, bracken and ivy – and three 'aliens' – rhododendron, sycamore and Himalayan balsam. He found that the locals caused four times more damage to local woodland biodiversity than the aliens. In other words, we have got it all wrong. Invaders are not a big cause of change in woodlands, and native invaders are far more of an issue than aliens.[19]

David Pearman, former president of the Botanical Society of the British Isles, edited a recent botanical atlas of the British Isles that collates 18 million records of 4,000 species. He says that despite the hype and fear, 'most aliens are rare, and occur overwhelmingly in and around towns and transport networks. They are generally uncommon in the semi-natural habitats that we most want to preserve.'[20] They like disturbed ground, whether graveyards or fields, herbaceous borders or railway cuttings. In such places, they keep nature going where other less hardy and less adventurous types falter. As Mabey puts it, 'many of them may be holding the bruised parts of the planet from falling apart'. If we don't like them, it may often be because we associate them with the messy things we do to nature, rather than for their inherent beastliness.

Britain gives the lie to the conservationists' mantra that aliens are self-evidently bad, and natives obviously more at home. In the real world, it is often impossible to tell the difference. Many aliens are so well integrated that they are assumed to be native. The snowdrop, usually the first wild flower to blossom in the new year, arrived from Brittany in the 16th century. It began as a garden flower, but soon went feral. It is much more widespread than the weeds we love to hate. But few in Britain know they are foreign, and nobody calls for them to be eradicated. Or consider the horse chestnut tree (*Aesculus hippocastanum*). It is the supplier of conkers to generations of schoolchildren and feted as the 'spreading chestnut tree' of a thousand rustic villages. As I write, newspapers report concern that an alien moth is attacking 'British' conker trees.[21] Nobody has noticed that the tree actually comes from the Balkans.

Meanwhile, some natives are assumed to be aliens. The sycamore (*Acer pseudoplatanus*) is widely reviled. Its seedlings and leaves smother garden lawns. Its reputation is likely to sink further if botanists are right that it is set to step into the gaps in woodlands created as ash trees succumb to ash dieback disease.[22] But the common presumption that it arrived in Britain only around 500 years ago does not hold up. It seems to have been widespread in Roman times.[23] John Rodwell of Lancaster University, who developed the British National Vegetation Classification, now argues that it has been a native all along.

The more researchers look, the more helpful many of the alien squatters appear. Buddleia (*Buddleia davidii*), brought from China by French botanists a century ago, is known as the 'butterfly bush', because its big honey-scented purple flowers provide superior nectar for native butterflies. It is one reason butterflies are leaving the British countryside, where wild flowers are scarce, and heading for the nectar-rich urban badlands where buddleia has taken root.[24] Britain's much-loved native red squirrel (*Sciurus vulgaris*) likes conifer plantations of non-native Norway spruce, Douglas fir and larch. Meanwhile, the red squirrel's alien American competitor, the grey squirrel (*Sciurus carolinensis*), prefers the native deciduous woodlands. There was controversy when the Countryside Council for Wales decided to remove alien conifers, the last local refuge for red squirrels. Under such circumstances, the legal presumption that alien species are bad and must be kept out looks misguided. Which should take precedence, the red squirrel or the conifer?

With all the headline-grabbing horror stories about Japanese knotweed and the rest, it is sometimes hard to remember that most of the 2,300 or so alien species in England are benign, and generally add colour and variety to the landscape. Thanks to them, the biodiversity of the British Isles is probably greater than it has ever been. 'No endemic is remotely threatened by any aliens', says Dickson. 'From a conservation perspective, I don't think they do much harm.' But our attitude to aliens is emotional, he says. It harks back to a 'merrie England' when everything was wonderful – 'a time many think we need to get back to'.

Cambridge geographer Stephen Trudgill says that our strong sense of what plants should be where is unhistorical. There is no true British flora, he says. Go back 8,000 years, to before we took up farming, and there was no golden age of stable, rich ecosystems full of native species. Instead the land was experiencing huge changes as it warmed after being covered in ice during the long ice age. Almost the entire flora and fauna of Britain has arrived in the past 10,000 years.[25]

Everything is visiting. Nothing is native. For Pearman, 'that makes the main reason for fearing the introduction of alien species [to Britain] invalid'. So why not celebrate the stories of their arrival? The seeds of hoary cress were transported in straw-filled mattresses that carried home the wounded from the Napoleonic wars two centuries ago, and were subsequently spread when the straw was given to a local farmer. A taxidermist's stuffed bird conveyed the hairy-stemmed Canadian fleabane from North America in the 17th century. A creeper called ivy-leaved toadflax arrived in Oxford around the same time in the packaging of marble statues looted from Italian houses.[26] The Romans brought devil's guts to relieve gout.

*

Fear of alien plants can turn to hatred of alien animals, especially recent arrivals. The signal crayfish (*Pacifastacus leniusculus*) was brought to Britain from North America in the 1970s to be farmed. It is bigger and tastier than native crayfish. But it is also more aggressive, breeds earlier, lays more eggs, tolerates dirtier water, eats a wider variety of food and carries a fungus that latches onto locals but to which it is itself immune. It can walk on land and burrow into river banks, causing them to collapse. With that skill set, it may be no surprise that it soon escaped and set about colonizing its new home. In little more than three decades, it has virtually supplanted native crayfish in lowland England, and by 2006 had made it onto a UK list of 'most wanted foreign species' compiled by the UK government's Environment Agency. Wanted as in 'wanted dead'.[27]

Succumbing to crude xenophobia, the Agency says the crayfish have 'taken advantage of Britain's welcoming living conditions' and 'overstayed their environmental visa'. Their 'crimes' include 'out-muscling native competition and spreading disease'. Echoing a complaint first made by British males about US forces stationed in Europe during the Second World War, they are 'oversized, over-sexed and over here'.[28] These remarks may have been made partly in jest – or at any rate concocted by a press officer eager to capture tabloid headlines. But they were only one step removed from the racism of the British National Party, which in a rare foray into ecology called the signal crayfish 'the Mike Tyson of crayfish ... a diseased, psychotic, evil, illegal immigrant colonist [that] totally devastates the indigenous environment'.[29] As with another fast-spreading, river-bank-burrowing but tasty alien crustacean, the Chinese mitten crab (*Eriocheir sinensis*), the best solution might just be to eat more of it. Celebrity chef Gordon Ramsay is a gourmet fan of the crab.[30]

Alien creatures seem to bring out the worst in even the most apparently animal-friendly people. 'There is a war going on in the parks, ponds, rivers and greenhouses of Britain', wrote Lucy Siegle in her ethical-living column in the *Observer* newspaper. It is a war she evidently approves of. She served up a photo-fit list of 'our top 10 unwanted non-native invasive species', with the top two places occupied by the signal crayfish and the grey squirrel.[31]

Many of the species appeared for no other reason than that they were 'fast-growing, aggressive and voracious'. They certainly didn't get the benefit of the doubt. The muntjac deer (*Muntiacus reevesi*) from China made number three even though 'scientists are not yet sure about the extent of the small deer's impact'. The worst crime of ring-necked parakeets, at number four, was that they represented 'potential competition for nest sites with some native birds'. The terrapin, released after the craze created by the TV hit *Teenage Mutant Ninja Turtles* in the 1980s, made number five even though 'we do not know the extent to which it is reproducing'. Actually, all the evidence is that British waters are too cold for them to breed at all.

Cuter alien animals may get a pass into public affection. The edible dormouse (*Glis glis*) is so named because, fattened up with almonds and chestnuts, it was a favourite snack food of the Romans. It was unknown in Britain until 1902, when a few of them were inducted to the private menagerie of Walter (later Lord) Rothschild in Hertfordshire. From there, some escaped and have since been tracked, almost village by village, on a steady odyssey across the Chiltern Hills of south-east England. They now number upwards of 10,000 and, most unusually for an alien species, have protection under wildlife legislation passed in 1981.[32] Not everyone is happy at this legislative anomaly. As the ever-grumpy *Daily Mail* put it in 2006: 'They live in your loft, breed like rabbits, go bump in the night, gnaw through wiring, strip fruit trees in your garden – and you can't touch them because they're a protected species.'[33] But the edible dormouse is nonetheless an honorary Brit.

The European rabbit appears in many places round the world, but is endangered in its native Iberia.[34] In fact, its spread may have been its making. The first rabbits to show up in Britain a millennium ago were feeble creatures, unable even to dig burrows. They were essentially domesticated animals, kept for food and fur. King Henry VIII had people dig burrows for them.[35] But along the way, they shaped up. In the 19th century, when their commercial value began to fade, the new improved, self-reliant version was let loose across the English countryside.

Another naturalized alien of longer standing is the house mouse, which hails from the Middle East and probably reached Britain with the Romans. Fallow deer came with William the Conqueror in the 11th century. That was also about when rabbits arrived from Iberia. Both were introduced for the same purpose, to provide a source of meat on poor heaths. Britons also seem relaxed about common pheasants, brought by the Saxons under Edward the Confessor and highly valued since Tudor times. Many 'native' birds are not so native. The little owl (*Athene noctua*), for instance, first arrived from Italy in the 19th century in the luggage of the eccentric traveller Charles Waterton. He bought a dozen in a bird market at the Pantheon in Rome. Five made it back to his estate in Yorkshire, where he released them. It is not clear if they survived, but a later flock set free in Kent by ornithologist Edmund Meade-Waldo in 1874 spread widely and have been at home ever since.[36]

The distinction between alien and native, and between domesticated and wild, is perhaps most confusing with wild boar (*Sus scrofa*). This is the only species known to have become extinct in Britain, but then to have returned as a free-roaming creature. In the Middle Ages there were a million or more boar rooting around the woodlands of England. They disappeared from the wild some 500 years ago, and from managed estates in the 18th century. But when their meat became popular again in the 1980s, foreign animals were brought in to stock new boar farms. There were soon escapes. An especially expansive wild population was established in south-east England after the Great Storm of 1987 destroyed the fences of a boar farm near Tenterden in Kent.[37]

The new wild boar has had a bad press. Some see it as an alien that should be repelled. In 2014, the British government began culling a herd of around 800 animals rampaging through gardens, panicking horses and rooting up crops around the Forest of Dean in Gloucestershire, on the grounds that they could be carrying diseases.[38] But is it an alien or a returning old English native? In law this matters a lot. Aliens are to be expunged, but the return of natives is a policy priority. Maybe the robins in the countryside have got it right. Robins follow the rooting boar around as assiduously as they do human gardeners digging the soil, and for similar reasons. Environment journalist George Monbiot says 'boars are the untidiest animals to have lived in this country since the Ice Age. This should commend them to anyone with an interest in the natural world.'[39] I agree. But Britons adopt what they like as their own. So some shed a tear for the recent demise of the red-necked wallabies (*Macropus rufogriseus*). This bit of marsupial Australia was introduced in the 1930s by Captain Henry Brocklehurst on his ancestral lands in Staffordshire. It thrived for half a century. The exiles from Tasmania munched away at the heather in the wilds of the Peak District, braving cold winters as their numbers declined.[40] The last died in 2009.[41]

Our feelings change, however. Prince Charles may echo the feelings of most today, when he says the red squirrel 'is one of the most utterly charming and irresistible of British native mammals'. But it wasn't always so. Once, the native red squirrel was widely regarded as a pest. The Scottish Highland Squirrel Club was dedicated to exterminating the 'devastating menace'. There was a bounty, and tails were used as a proof of kill. Between 1903 and 1946, the club's members killed more than 100,000 on the Beaufort, Cawdor and other estates.[42] Back then, many preferred the grey squirrel, an invader from North America busy expelling its red cousin. The newcomer was greeted with excitement after releases in the late 19th century by the Duke of Bedford from his Woburn Park estate, and by Thomas Brocklehurst, a Victorian banker and ancestor of wallaby-releasing Henry Brocklehurst. The *Manchester Guardian* declared that the American grey 'seems wishful to becoming a colonist in Britain. Many of us will welcome it, for it is a sociable, easily tamed animal, with attractive ways.'[43]

Like alien plants, alien animals often provide useful services for domestic species. Such as food. Common buzzards, red kites and other native birds of prey making a comeback across the British skies in recent years are partial to alien rabbits. In fact, some argue that it was the loss of rabbits to myxomatosis in the mid-20th century that caused the decline of these birds of prey in the first place. The same may also go for Scottish wildcats and polecats, which grew scarce around the same time.

Many of those invading species accused of pushing out natives have simply moved into ecological space created by the decline of the natives. This is a recurrent theme in the story of aliens. The American mink (*Neovison vison*) spread widely across Britain in the mid-20th century, after escaping from hundreds of fur farms. It was widely blamed for the collapse of the number of otters (*Lutra lutra*) along British rivers during the same period, the argument being that it out-competed the otters for food. But in the past twenty years, otters have made a strong comeback – at the expense of mink. It begins to look more likely that the mink simply took advantage of the otter's decline, rather than causing it. Something else saw off the otter – mostly likely agricultural pesticides. As the worst of the chemicals have been withdrawn, the otters have returned.[44]

Deer too create a problem for our simple distinctions between good natives and bad aliens. The native red deer (*Cervus elaphus*) has become a major pest across Scotland. An adaptable animal, it has thrived too well since its former woodland habitat across Scotland was erased. Now its grazing prevents new native saplings growing. Meanwhile, Britain's population of Chinese water deer (*Hydropotes inermis*) is of growing global importance. Its British population began as an import in the 1890s by the Duke of Bedford at Woburn Park. Some of the duke's herd escaped into the English countryside in the 1940s. Since then, its numbers have dwindled back home along the River Yangtze. Of the 10,000 believed left in the wild, Britain has a thousand. (Another outlier population inhabits the demilitarized zone between North and South Korea.)[45]

*

CABI reported to the British government in 2010 that the total cost to the British economy of alien invaders was £1.7 billion a year. That was the same report that put the 'cost' of Japanese knotweed at £170 million.[46] It totted up everything from pest damage to crops to the increased risk of flooding if coastal defences are attacked by a boring beetle; from rabbit damage to lost recreational value. More than 60 per cent is accounted for by foreign agricultural pests, the cost of weed control and other agricultural losses. Britain's 40 million rabbits inflict a fifth of the bill, half from the cost of control and half an estimate of the value of nibbled pasture grass and grains. Other costs include collisions between deer and cars, eradicating rhododendrons in national parks, squirrel damage in woodlands, and parasites in salmon farms.

Dick Shaw says this was 'the most thorough assessment ever done in any country'. And that may be true. 'We expected criticism but had very little', Shaw told me. That is surprising. The figure is certainly largely uncontested in official circles, even among those whose job is to question such things. When, in early 2014, the UK Parliament's Environmental Audit committee launched an inquiry into Britain's approach to invasive species, it never thought to ask if the CABI figure was open to question.

From my own reading, some of its costings seem plausible, though they rather stretch the definition of what is genuinely alien. Maybe every rabbit does cost the British economy £5, though after a thousand years perhaps we should stop classing it as an outsider. Other costs seem to arise from eradication programmes caused by misplaced fears about alien species. But others again look to me distinctly far-fetched, and not real costs at all. The supposed bill for Japanese knotweed is one, as we have seen. Another is the cost of floating pennywort (*Hydrocotyle ranunculoides*), an American aquatic weed that has been spreading across the surface of British waterways, and for which Shaw has a weevil biocontrol on trial. You might imagine that the 'cost' of this weed – £6.1 million – would be based on actual money spent keeping waterways clear of the weed. But that figure is only around half a million pounds.

CABI has instead produced a figure ten times higher that is an assessment of lost recreational value. This is in turn based on an opinion poll that asked people to put a cash value on their 'willingness to pay' to visit the nation's canals and rivers. I cannot imagine how I would answer such a question, and certainly wouldn't want government policy based on my answer. But even if you believe the answers, surely most canal-goers who found their favourite bit of waterway to be blanketed in weed would simply choose another. In what sense has the national economy been damaged?

For some people, these kinds of statistics help underpin an indiscriminate fear of foreigners that I find distressing and perverse. So I was interested to find that the Royal Society for the Protection of Birds now often takes a benign view of new arrivals – perhaps because many of its million members like nothing more than a new bird to go and watch. Britain is a happy hunting ground for new bird species, often because of the attractions of changing climate. Twenty new wetland species have arrived since 1960 alone, including the little egret, which first bred in Britain in 1996, the spoonbill in 1998, the pectoral sandpiper in 2004, the cattle egret in 2008, the purple heron in 2010, and the great white egret in 2012. 'We are not against all exotics', said the RSPB's Grahame Madge. 'We have many colonists arriving from Europe, partly probably in response to climate change. We won't introduce them or help them, but if they come then we will ensure that they are at home here.' They won't be taking shotguns into their protected areas, he says. 'The birds that have established here have been relatively benign. The little egret, for instance, is a favourite now.'

PART TWO
MYTHS AND DEMONS

Alien species encourage myth-making. We all love a villain, but the vilification of aliens makes for counter-productive conservation.

6
Ecological cleansing

Many conservationists see themselves as engaged in a war against alien species. To protect natural ecosystems, the outsiders must be shot, poisoned, burned, hunted by dogs, uprooted, chainsawed or gobbled up by specially introduced insects. Many make their cases on grounds of protecting some vital national resource, whether crops or forests or water supplies. There can be no compromises. No talk of animal rights. Not even any regard for the local biodiversity that might be lost as the foreigners are removed. All methods are acceptable in the battle to purify encroached ecosystems and recreate the pristine. Because aliens are bad, and natives are good.

Few countries have gone on the offensive against alien species with the zeal of post-apartheid South Africa. It has deployed more than 25,000 people every year for almost two decades to rip up whole landscapes and root out 68 species of alien trees, most of them planted during colonial and apartheid eras. The foreign trees are, says the head of the programme Guy Preston, 'analogous to a cancer' in the country.[1] The rhetoric suggests a desire to purge the legacy of foreign species along with that of foreign occupation and apartheid politics. But has the rhetoric got in the way of practical policy and good science? Why are there more invasive species than ever across the veldt?

European settlers brought in a range of exotic trees in the late 19th century to, as they saw it, 'improve' the land, and make the bare grasslands look more like home. Back then, the trees were regarded as superior to native species, because they were faster-growing and said to be beneficial to the wider environment.

An Australian acacia called black wattle (*Acacia mearnsii*), for instance, stabilized sand dunes, while producing charcoal, firewood, tannin for the leather industry, and timber to shore up shafts in the gold mines. But the foreign trees often spread to places they were not wanted, in particular invading farmland. Government officials today claim that 20 million hectares of the country have been covered by foreign acacia, pines, eucalyptus, mesquite, prickly pear, water hyacinth, giant reeds and others. The official policy is that they have to be removed.

The purge has been masterminded in part by ecologists fearful of foreign trees invading the *fynbos* district in the Cape region. This is one of the most biodiverse spots on the planet with more than 8,000 species of plants, two-thirds of them found nowhere else. But the ecologists' pitch to government has been that the alien species could pose a threat to water supplies. Researchers claim that around 7 per cent of the nation's rainfall is taken by foreign trees. In future, they could increase their take by a third around water-scarce cities such as Cape Town.[2] In a country where only 10 per cent of rainfall makes it into its parched and seasonal rivers, there is understandable alarm. Willem de Lange and Brian van Wilgen of the Centre for Invasion Biology in Stellenbosch estimate that the water loss translates to an economic loss to the nation of around half a billion US dollars each year. Controlling invader species, researchers argue, is a more cost-effective way of maintaining water supplies than building dams.[3]

Faced with such advice, the post-apartheid government quickly decided to start eradicating the alien water-guzzlers.[4] A nationwide programme began in 1995 under the name Working for Water. It is a valuable job-creation scheme. And the backbreaking work is portrayed as heroic and patriotic. Poor unemployed people from the country's slums are recruited to abseil down cliffs, wade through wetlands, dangle from helicopters and brave snakes in their war on the aliens. The initiative has also enhanced South Africa's reputation as a world leader in fighting invasive species, helping it to secure the home for the headquarters of the IUCN's Global Invasive Species Programme, established in 1997 at the Kirstenbosch botanical gardens near Cape Town. Ironically, Kirstenbosch is the place from where many of the alien species either escaped or were deliberately distributed.

But does this alliance between conservation and job-creation deliver what it promises? Hydrologists point out that the figures about how much water is 'taken' by the invader trees may not be the same as the amount of water that would be 'saved' if the trees were removed. This is because, without the alien trees, other vegetation would be taking water. And if there were no vegetation then evaporation losses from the ground would soar.[5] Elsewhere in Africa, the same trees are being promoted by ecologists to protect soils that store water. For instance, the UN Food and Agriculture Organization and aid agencies have encouraged farmers in arid Africa to plant species blacklisted in South Africa, such as black wattle, to improve the water-holding capacity of soils and generate income for farmers. It is, says Christian Kull of Monash University in Melbourne, far from clear why those trees are regarded in such a different light when growing in South Africa.[6]

A second critical question is whether the massive project is eradicating the trees as intended. For several years, the South African government has been claiming that it has purged around 2 million hectares of aliens. It looks like progress. Except that, according to a 2012 assessment by some of the academic pioneers of Working for Water, the claim does not stand up. According to van Wilgen, despite the billions of dollars spent, the amount of land covered by the target species *increased* by more than 4 per cent between 1996 and 2008.[7]

A huge amount of hacking and spraying has gone on. A great deal of sweat has been expended. Some areas have been cleared, notably in Table Mountain National Park. But van Wilgen says that job-creation priorities have taken precedence over making sure the job gets done. 'The ability of Working for Water to provide employment has become its major attraction to politicians.' Meanwhile, there has been very little official monitoring of what the project has achieved, and in particular whether alien species have returned to cleared areas. It now appears, he says, that seeds persist in the soil and readily regrow after the work gangs have moved on, continuing their colonization. Often they invade new lands faster than other areas have been cleared.

*

Despite such setbacks, conservationists worldwide have grown increasingly ambitious in their plans to restore ecosystems through eradication schemes. Most attempts have concentrated on remote islands. These are places where there are some of the most visible signs of the presence of invaders. They are also places where eradication is much more achievable. Once got rid of, there is a fair chance that many aliens will stay away for good. But even on islands, such programmes are rarely so simple. The knock-on effects can cascade through ecosystems in unexpected ways. Sometimes the cure is worse than the disease.

You can't get anywhere much more remote than the British-administered island of South Georgia in the Southern Ocean near Antarctica. In January 2013, the island had unusual visitors. A team of reindeer herders from the Sami people of Lapland in northern Norway made a long trip south to corral and shoot the world's most southerly herd of reindeer (*Rangifer tarandus*). It was going to be a big job, over two austral summers. The island was overrun with almost 7,000 reindeer, all descendants of animals introduced to the chilly pastures a century ago as a handy local food source for passing Norwegian whalers. Whaling ceased in the 1960s. Since then, apart from during the island's brief capture by Argentine troops in 1982, South Georgia's only inhabitants have been visiting scientists and officials. Nobody had been eating the reindeer, and their numbers soared.

The reindeer were accused of trampling native plants, including the tussock grasses where burrowing seabirds nest. Conservationists fear that as the island's glaciers melt, reindeer numbers could grow further, dooming the birdlife. The island is home to colonies of indigenous pintail ducks, burrowing petrels and South Georgia pipits, the world's most southerly songbird. So, the Sami herded the reindeer into pens, where they were finished off with a gunshot to the head. And in areas where the terrain made that impossible, the animals were tracked down by Sami marksmen. By the end of the 2014 killing season, the last of the reindeer had been killed. Some of the carcases lay across the island, while others were cut up and sold to cruise ships, though many reindeer joints remained in cold store in the Falklands.[8]

But the reindeer are not the only target of the ecological restorers. To protect the bird colonies, they also reckon they have to eliminate the island's millions of brown rats (*Rattus norvegicus*), whose ancestors first scuttled ashore from the boats of 18th-century seal-clubbers. Like the reindeer they found no natural predators, but plenty of birds' eggs to eat. So, even as marksmen rounded up the reindeer, three more helicopters began criss-crossing the island, braving gale-force winds and blizzards to drop the first of 100 million pellets containing the anti-coagulant brodifacoum onto the tundra.

The four-year rat eradication is costing $10 million, paid for by Frederik Paulsen Jnr, a Swedish pharmaceuticals billionaire, keen ornithologist, polar explorer and founder of the South Georgia Heritage Trust. If successful, the South Georgia purges will be the one of the largest ever deliberate eradications of an alien population of animals. The aim, says project leader Tony Martin from the University of Dundee, is not just to maintain a fragile status quo, but to get back 'to the way things were before 1775', the year Captain James Cook claimed the island for Britain. It will allow millions of birds to return. For the programme to be a success, 'every single rat must be eradicated'. But that requires bombarding the island with some 270 tonnes of a persistent poison, with a half-life in soils of around six months. Some think that is a price not worth paying. The consequences are too uncertain. Ken Collins of Southampton University, a British specialist on island ecology, told me: 'They are flooding the island with poison. I am a bit open-jawed about that.'

An obvious concern is that birds might be poisoned too. The scientists involved admit this is a risk, particularly to the brown skua, which partly feeds on rat carcases, but also to endemic South Georgia pintails and sheathbills. Critics point out that when the same poison was tried on Rat Island in the Aleutians, the most conspicuous victims were 41 bald eagles that died after eating dead rats. 'Some non-target (bird) mortality is sadly inescapable, but losses will be recovered over the space of a small number of years', says the Trust, at the halfway stage of the operation.

Even if Paulsen's pogrom gets rid of the rats, there will still be the mice. Once thought rare on South Georgia, it now seems that they have all along been hunkering down in their burrows – albeit in small numbers where there are rats.[9] The rat bait will kill mice, but they are a much harder target, because they feed only in small areas and may miss the poison. So, as the rodent eradication plan drawn up by the British government's Overseas Territory Environment Programme noted, there is 'a risk that these populations will rapidly increase after the removal of rats'. In 2013, the rat eradication plan was augmented by a separately funded mouse killing trial.[10]

When I asked the Trust about all this, they admitted that mouse eradications in other places had only a 50 per cent success rate. But they were sanguine about the risks to seabirds from a boom in the mouse population, because 'there are no rats present in the areas where mice have been detected, so removing rats should have no effect on those mouse populations if the mouse eradication is unsuccessful'. Maybe, but equally this suggests that mice live where rats don't, and if the rats are removed while the mice survive, then the mice may spread. Given the alarming history of super-size mice eating their way through the birds of Gough Island, the saga may not yet be over.

*

The perils of such efforts to re-engineer even simple island ecosystems is shown starkly on Macquarie Island, another bleak, treeless speck of rock in the Southern Ocean, halfway between Australia and Antarctica. After being discovered by Europeans in 1810, it was regularly visited by hunters eager to slaughter its seals in pursuit of fur and blubber. It also had huge colonies of sea birds. Every royal penguin on Earth nests here. Thanks to the human visitors, however, the island soon stocked up on foreign animals.

First came rats, jumping from the sealers' ships. The rats ate food stores that the sealers left buried in wooden barrels. So the sealers brought in cats to hunt the rats, and then rabbits as a living larder to be taken for the pot during their visits. Despite being hunted by both humans and cats, the rabbits enjoyed the damp and chilly terrain rather too well. By the mid-20th century, there were some 150,000 of them, a thousand for every square kilometre of the island, nibbling their way through its tussock grasses.

By then, conservationists were in charge of the island. They decided something had to be done before the rabbits turned the place into desert. That something would be a flea, *Spilopsyllus cuniculi (Dale)*, that lives in the fur of rabbits and carries the myxomatosis virus.[11] Introduced in 1968, the virus cut the rabbit population by 90 per cent. The island's vegetation began to recover. But there was a problem: the cats. Without rabbits to eat, the cats turned their claws on the island's ground-nesting sea birds. On Macquarie, as elsewhere, birdlife is conservation's royalty. So, starting in 1985, the Australian government started shooting the cats. Many ecologists at the time thought this was a dumb move, and they were right. Fifteen years later, the island was cat-free. But without the cats, the rats ran riot, eating the birds in ever greater quantities. And the rabbits had never gone away. Without the cats, their numbers also revived dramatically. Their renewed nibbling of the grass began to cause soil erosion, culminating in a landslip in 2006 that buried alive an important penguin and albatross colony.[12]

It looked like it was back to square one.[13] So in 2007, the Australian government in Canberra hatched a grand plan, costing A$24 million. Everything had to go. They would eradicate rats, rabbits and the increasing numbers of mice by mass poisoning using brodifacoum, with any survivors taken out by follow-up teams of hunting dogs.[14]

Sometime in 2012 the dogs finally nailed what were thought to be the last thirteen rabbits. Job done? Scientists back in Australia, sitting at their screens running computer models of the island's ecosystem, believe so. But don't hold your breath. For the poison seeped into the soils and vegetation and began killing hundreds of the birds, including kelp gulls, giant petrels, black ducks and skuas.[15] The story seems far from over. Nobody yet knows if the eradications have been achieved.[16] And if rabbits start popping their heads out of their burrows again, there may soon be a call to bring the cats back.

*

Despite such debacles, billions of dollars have been spent trying to eliminate alien species.[17] As we saw in Chapter 3, blitzing a few acres of *Caulerpa taxifolia* from a bay in California a decade ago cost $6 million. And there is also an alarming failure rate. As we will see in Chapter 10, of 43 projects to eradicate or control aliens on the Galapagos Islands in the Pacific, only nine were successful. Globally, most successful eradications have involved removing rats on small islands. Victories elsewhere are rare.

Many expensive eradications simply prove unnecessary. The American muskrat (*Ondatra zibethica*) originally arrived in Europe courtesy of Joseph II, the Prince Colloredo-Mannsfeld, who in 1905 went on a hunting expedition to Alaska and brought back five of the furry river rodents. He released them onto his estate near Prague, in what is now the Czech Republic, and farmed them for their fur. He also bred them for sale to other landowners. The inevitable escapes soon created wild populations in many places. There was consternation. The *New York Times* reported that 'like the rabbit in Australia and the English sparrow in America, the muskrat has developed a long list of evil traits, scoffing river crabs and undermining river banks'.[18] Many governments set up eradication programmes. But almost all failed. The animals continued to spread. They can now be found from the western shores of France and northern Scandinavia to the Russian Far East and Japan.

Soon the panic died down, however. As frequently happens with alien invasions, muskrat numbers boomed, peaked and then declined. Muskrats have turned out not to be too troublesome after all. Yes, they are distrusted in the Netherlands for making holes in dykes, and elsewhere for gnawing the linings of suburban swimming pools. But they have added to local biodiversity and slipped out of most lists of dangerous aliens. In 2012, a study of alien species by the European Environment Agency ignored them altogether, apart from a passing thank-you to them for eating invasive zebra mussels. A century on, the muskrat is almost at home.

Many ostensibly successful efforts to eradicate incomers have suffered unintended consequences that are more disruptive than the original invasions. That happened with the British effort in the 1930s to trap and kill its own population of escaped muskrat. The effort was, formally speaking, a success. Every last animal was exterminated. But the traps also slaughtered their close relative, the native and much-loved water vole (*Arvicola terrestris*). According to a later review by zoologist Charles Elton, the traps killed more than twice as many water voles as muskrats, contributing to a catastrophic decline of the water vole in the mid-20th century.[19] Ironically, the decline has subsequently, and erroneously, been blamed on predation by another escaped furry American introduction, the mink. Thus one alien was blamed for an ecological mishap that was probably due to earlier efforts to exterminate another.[20]

The law of unintended consequences held too on Sarigan island, a volcanic island in the Western Pacific near Guam. This former German penal colony was taken over by the US after the Second World War and its human residents all removed. In the 1990s, the US-appointed administrators turned the depopulated island into a nature reserve, with the aim of rehabilitating forest habitat for native species like the secretive and endangered local fowl, the Micronesian megapode. As a first step, another forced removal was required. The feral goats and pigs had to go.[21] But instead of native forest blossoming, the place was overrun by an alien vine from other Pacific islands called *Operculina ventricosa*. It had previously been the goats' main food, yet was grazed so well that botanists failed to notice its presence when the goat eradication was being planned. But it has now run riot and covers most of the island.[22]

The tangled ecology of San Francisco Bay also tripped up would-be ecological restorers. In the mid-20th century, engineers had drained many of the bay's marshes and mud banks for building projects. But attitudes changed. Conservationists became concerned about the loss of natural habitat, and from the 1970s, engineers spent more millions of dollars on plugging up their drains to restore lost mudflats, salt marshes and other wetlands.[23] As part of this programme, the Army Corps of Engineers began planting re-wetted marshes with a cordgrass native to the eastern US (*Spartina alterniflora*).

This new grass began to interbreed with its close relative, the local California cordgrass (*Spartina foliosa*). The result was a new hybrid grass that colonized much more aggressively than either of its forebears. It spread to areas no one had intended, blanketing previously open mudflats, clogging channels, getting in the way of the oyster farmers and – worst of all, for many – spoiling million-dollar views and damaging the value of upmarket waterfront properties.[24] So a decade ago, the bay authorities launched a multi-million dollar project to rid the bay of both the alien from the east and the hybrid.

But that went wrong, too. It turned out that one of the bay's most totemic and endangered birds, the chicken-sized and largely flightless California clapper rail (*Rallus longirostris obsoletus*), had grown partial to the new hybrid grass. The grass grew more densely than the local version, and didn't die back during the winter, so providing better cover and nesting habitat for the secretive bird. During the 1990s, as the hybrid spread, the rail population had soared. But after 2004, as the eradication programme got under way, the bird's numbers crashed. There was no mistaking the cause. In time and space, the bird population declined following the eradication of the alien grass.

Conservationists now face a dilemma. Should they carry on with the eradication and hope the birds eventually recover? Or should they walk away? In 2013, nobody seemed sure what to do for the best.[25] The website of the California-based bird conservation group Point Blue Conservation Science still backed removing the grass, saying 'invasive non-native plant species' are a primary threat to the rail, and that the hybrid 'may reduce channel and mudflat areas important for foraging rails'.[26] But bay managers had banned further removal where there are nesting rails. The rail must be as confused as the rest of us.

*

Eradication can take many forms. In South Africa, ridding the land of alien trees and shrubs has generally involved chainsaws and machetes and human sweat. Muskrats faced traps. Reindeer were shot. Rats and other rodents generally suffer bombardment with brodifacoum or some other poison. But an increasingly popular idea is deploying biological weapons such as insects or diseases. Dick Shaw at CABI, Britain's chief bio-control practitioners, reckons there have been some 140 releases of bio-controls in Europe alone.

The idea is simple. Aliens often prosper because they come alone, unencumbered by the predators and other natural enemies they encounter back home. Nothing in their new environment has evolved, or has yet learned how, to eat them, or inflict diseases on them. In most cases this does not secure their success. Something gets them. Few predator species will forgo a free meal, even if the prey is novel. By the same token, the new arrivals have no honed defences against what they may find in their new home. So the stakes are high in the early days. But if the newcomers do not succumb to some native species that takes a liking to them, then they can have a free run for a while. Under those circumstances, if you want to be rid of them, the logical way to fight them is to track down and import a predator or some disease from their homeland as their nemesis in the new land. Set an alien to catch an alien.

Among practitioners, Australians are regarded as the masters. Their reputation lies mostly with eradicating the prickly pear cactus (*Opuntia monacantha*), an introduction that goes back as long as English settlers in Australia. When Captain Arthur Phillip was on his way to set up the first European settlement at Botany Bay in 1788, he stopped off in Brazil to pick up some prickly pears. He wanted the cochineal insects (*Dactylopius coccus*) that live on them. When squashed, they yield a crimson dye then used in British military tunics. Until then, rather embarrassingly, the British were dependent for the dye on the Spanish and Portuguese, who held a world monopoly through their American colonies. But while the prickly pear enjoyed life in the new British colony, the cochineal insect did not. The result was no dye but a spreading plague of cacti. The plague persisted and grew for more than a century until government scientists brought it under control in the 1920s by releasing an Argentine moth that eats the cactus.[27]

One danger with introduced bio-controls is that they become too much at home, and start eating or infecting the natives as well. Rather than re-establishing ecological peace, they can become a problem far more difficult to handle, and widespread, than the original alien. That happened when the Argentine cactus moth was tried against prickly pears on a handful of Caribbean islands. It has since spread unbidden into the US. But the classic case – again in Australia – is the cane toad. Fresh from success in tackling prickly pear, Australia's bio-controllers decided to take on native beetles destroying the sugar cane fields of Queensland. In 1935, they brought in a giant South American toad (*Bufo marinus*) to eat them. The toad had done the same job successfully in the cane fields of Hawaii, the Philippines and Puerto Rico. But when Queenslanders released about 60,000 of them, the toads largely ignored the beetles, hopped out of the cane fields and explored their new homeland. They developed a taste for a wide variety of Australian insects and other invertebrates.

Weighing a kilogram or more, and measuring up to 15 centimetres long, the cane toads were bigger than the average toad. But the real problem was that they had a gland on the back of their heads that was highly poisonous to almost anything that tried to eat them. In the toads' homeland, potential predators had learned to be wary and the glands acted as a deterrent. In Australia, they made a lethal meal for passing predators who knew no better, like snakes and even freshwater crocodiles.

The innocent locals were massacred in large numbers. The carnage spread as, determined to make the most of their freedom, the cane toads developed longer legs and headed west.

Top Australian naturalist Tim Flannery remembers: 'A friend of mine was camping on a river in western Queensland when the toads went through. He went out one evening to fish, and was distracted by a nauseating smell. Following it upstream he discovered a logjam of dead crocodiles … a single toad was enough to kill even the largest of them.'

By the 1980s, the toads had hopped across state lines into the Northern Territories, headed for Kakadu national park, Australia's wildest place. Fears grew of an ecological Armageddon. 'Kakadu is lost', Mike Tyler of Adelaide University told a breathless media in 2002. 'They're going to be more prolific here than anywhere else, and they're going to get bigger. The cane toad will become the dominant life form in a little bit of Australia that we thought was pristine.' The predictions seemed to be coming true.[28]

Rick Shine of the University of Sydney recalls: 'As the toads swept through Kakadu, we saw over 90 per cent mortality in large predators like varanid lizards, blue-tongued skinks, freshwater crocs and some of the big snakes.' The nation was on toad alert. State governments spent some A$20 million in an effort to halt the toad's advance. Police officers patrolled the borders of Western Australia, using mirrors on sticks to check for toads hiding under cars, in the way forces in war zones check for explosives. 'Rarely has an invasive animal been so widely reviled by the general populace', says Shine. Community groups sprang up to fight the invasion. Websites like www.stopthetoad.org.au proliferated. Scientists were caught up in the panic. 'A research programme on toad impacts is more likely to be funded if toads are viewed as catastrophic', Shine noted later in a sardonic paper on what he came to see as a hysterical over-reaction.[29]

All efforts to halt the toads' takeover failed. By 2009, they had reached Western Australia. They now occupy more than a million square kilometres across the north of the country. And yet, says Shine, they are not proving as bad as once feared. The beasts of the outback started getting the measure of the toxic newcomer. 'The vast majority of smaller predators rapidly learned not to eat the toxic toads', says Shine. And slowly the bigger predators got the message, too. Freshwater crocs learned to nibble the toads' back legs and leave the rest. The black kite and a crow have both developed the trick of avoiding the poison gland by attacking its belly or throat. 'Aversion learning', Shine calls it.

Evolution also seems to be working to defeat the new menace. Some snake species developed tolerance to the toxin. Researchers found that within a few generations, blacksnakes from toad-colonized areas had developed smaller heads, since small-headed snakes can eat only smaller, less toxic toads. After initial declines as the cane toads advanced, there are revivals among monitor lizards, crocodiles and other species, including a local cat-sized marsupial called a quoll. Kakadu is not lost. 'Our predictions were dramatically in error', Shine reported in 2011. 'No native species have gone extinct.' Birds and rodents turn out to tolerate the toxin.[30] Even humans are adapting. In Queensland, cane toads have become part of the culture. Initially it was open season, with cane toad golf especially popular. But the gutsy newcomer is now rather admired. The National Trust of Queensland considered it for a list of state icons; cane toad purses and other souvenirs are rife; the state rugby league team is known as the Cane Toads; and outsiders have even taken to calling Queenslanders 'cane toads'. But it has been a long haul.

*

It would be foolish to claim that alien species never do any harm, or that efforts to uproot them are always doomed to fail. Neither is true. And advances in techniques may improve the chances of success, especially for bio-controls, where the main trick is to ensure that the bio-control won't start eating its way through native species. But we need a sense of proportion. And too often the species police forget that. From Macquarie to South Africa, a misplaced belief that ecosystems can be brought back to equilibrium by removing one presumed felon has created new mayhem.

Take an area of the Florida Everglades known locally as the hole-in-the-donut. This area of farmland was established in the early 20th century across 10,000 hectares of slightly higher ground at the heart of the Everglades. The new farms were seen as the prototypes for a large-scale drainage of the swamps being pursued by the then governor. But ideas changed. Once seen as a putrid marsh full of dangerous creatures and disease, the Everglades was gradually rebranded as a biodiverse wetland worthy of protection.

In 1947, the undrained area – the donut surrounding the hole – was turned into the Everglades National Park. Later, park authorities decided to pursue their conservation mission by filling the hole in their donut. In the 1970s they bought out the farmers and waited for the marsh vegetation to reclaim the abandoned fields. But the ploughed, drained and heavily fertilized soils were now very unlike the rest of the park. And the nature that showed up mostly comprised non-native species, like the Brazilian pepper-tree. 'Every weed in South Florida found it to be a great place to grow', says Jack Ewel, a restoration biologist who worked in Florida for many years but is now at the University of North Carolina.[31]

That alien invasion was not in the script. Nor was the fact that the thick alien undergrowth swiftly became a haven for much local wildlife. Among those that liked it and dropped by with increasing frequency were the last hundred or so of the Florida panther, an endangered local sub-species of cougar (*Puma concolor coryi*). Fleeing traffic and tourism in the more picturesque parts of the park, the recently-installed state animal found the overgrown farms in the 'hole' an ideal refuge. Its numbers began to rise.

But, panther habitat or not, the revegetated hole was not the pristine wetland that the parks people wanted. A pragmatic approach would have been to leave the hole as it was. But after two decades of failed efforts to root out the 'weeds', the park authorities decided on a final solution. They would rip out all the vegetation, strip the soil back to bare rock and start again from scratch. 'I was incredulous', says Ewel. Giant scraping machines removed some 7 million tonnes of soil and dumped it in mounds that remain to this day. It worked, after a fashion. Eradicating every last trace of the invasive vegetation did allow some native wetland plants to colonize the hole. But there is an unanswered question about what to do with the 25,000 dump-truck loads of soil piled up in huge mounds.

And what about the panthers? Concerned about the possibility that the ecological restoration might result in their extinction, park authorities augmented their stock with some females from Texas. The newcomers are a different sub-species, but a good enough match for breeding. There are now 160 panthers in Florida, a mix of the Florida and Texas sub-species and a growing number of hybrids. Some taxonomists talk of deleting the distinction between the sub-species altogether. Ewel is not impressed. 'Florida's state mammal is now a novel hybrid', he told me. A sub-species of an iconic cat has been sacrificed in order to fill the hole-in-the-donut.

Was it worth it? Only if you think there is no value in unique ecosystems such as the rich panther habitat created where the farms had been. Only if you think that spending tens of millions of dollars on trying to put back your notion of the pristine is conservation money well spent. And only if you forget about the disappeared sub-species. Only, perhaps, if you have no sense of proportion.

That valuable commodity often seems to go missing where alien species are concerned. Ecologists and their backers are overcome by a cleansing zeal. The most cock-eyed story of alien extermination I have come across is the martyrdom of the ruddy duck. Birdwatchers don't often condone the shooting of birds.

And they are usually rather indulgent about foreign flocks, perhaps for the sheer joy of seeing something new through their binoculars. But the ruddy duck, though it has been resident in Britain for the past half century, is different. *Oxyura jamaicensis* is a North American bird, one of a group known as stiff-tailed ducks. It is larger than the average duck. The story is all very embarrassing for conservationists.

The first ruddy ducks arrived in Europe at the invitation of one of the 20th century's most famous ornithologists and conservationists. Sir Peter Scott was a friend of royals, founder of WWF, star of TV wildlife programmes and knight of the realm – the Sir David Attenborough of his day. His pride and joy was the Slimbridge reserve for water birds, which he established on soggy sugar-beet fields beside the River Severn in the English West Country. In 1948, soon after the reserve opened, he shipped in three pairs of ruddy ducks from a collector in Salt Lake City. Scott normally clipped the wings of foreign birds at Slimbridge, to prevent their escape. But these ducks proved hard to catch and several dozen of the chestnut-breasted new arrivals evaded the clippers and flew off. By 1960, they were nesting and breeding widely in reed beds in the surrounding countryside. They did no obvious harm, and fanned out across Britain.

But alarm bells rang when they crossed the English Channel and began heading south, looking for new nesting sites. Suddenly it looked like an American invasion. Worse, having reached Spain in the early 1980s, a few of them interbred with the local and endangered white-headed duck (*Oxyura leucocephala*). This species was down to its last 10,000 individuals, a quarter of them in Spain. In 1991, the interbreeding produced the first fertile hybrid. More followed.[32]

Spanish ornithologists were proud of having brought the white-headed duck back from near-extinction in the 1970s. Now they feared losing its genetic purity. They began shooting both the hybrids and the migrant ruddy ducks, and launched a public campaign to get the rest of Europe to cull the ruddy ducks to prevent further contamination from American duck genes. Ramón Martín of the Spanish Ornithological Society upped the anti-American sentiment by describing 'gang rape' when the ruddy ducks descended on the native señoritas.[33]

The ruddy duck became an international incident. Eventually, Britain and other European countries agreed to kill off all of the thousands of ruddy ducks in Europe, shooting them out of the skies and pouring paraffin onto eggs in their nests. The Royal Society for the Protection of Birds – often the arbiter of British public policy on birds – gave its support, and the British shooting began in 1999. By 2012, spending on hired guns had exceeded £5 million, and some 6,500 ducks had been downed at more than a hundred breeding and wintering sites. By the summer of 2014, just ten females remained, and the government vowed to complete the extermination by the end of 2015, even though flushing out the last would cost around £3,000 a bird, making them some of the most expensive ducks in the world.

France has also been shooting the interlopers, but the Netherlands has refused. Many question the wisdom of all this. The fast-living, fast-breeding American ducks, hiding in their reed beds, may well survive efforts to exterminate them. They are now in at least nine European countries. In any case, it is not clear how extensive the interbreeding is, and what threat it actually poses to the survival of pure-bred white-tailed ducks. Some naturalists say the whole idea of pure-bred ducks is an illusion. Cross-breeding between species, and hybridization, are routine among waterfowl. They do it all the time. So why make an issue out of the involvement of ruddy ducks?[34]

The battle of the ruddy duck is a fitting set-piece in the war over how we should see nature and what protecting species should be about. Conservationists, it seems, are dedicated to protecting the weak and vulnerable, the endangered and the abused. Nature generally promotes the strong and the wily, the resilient and versatile. Conservationists support the pure-bred white-tailed duck. Nature backs the ruddy duck and its bastard offspring.

7
Myths of the Aliens

The idea that alien species are things to be feared is relatively recent. Victorians and their Acclimatization Societies had no such concerns. But as scientists developed ideas about ecosystems as tightly knit associations of species that had evolved together to create a unified whole, the conclusion that outsiders are 'other' and malign was almost inevitable. The sense that aliens represented a global scourge was first given full scientific expression by British zoologist and ecologist Charles Elton. He is the founder of what academics today call invasion biology.

A mild-mannered and unpretentious Oxford academic with a balding head and round glasses, Elton hated conferences and committees and preferred working alone in the field. But he was also something of a pugilist, both as an amateur boxer and a scientific iconoclast. Elton made his name in the 1920s researching Norwegian lemmings (*Lemmus lemmus*). He recorded how these small Arctic rodents reproduced rapidly until they ran out of grass. They then went on mass migrations, sometimes swimming huge rivers in their efforts to find food. Often they drowned. This discovery made him the originator of our favourite myth about how lemmings commit mass suicide.[1]

During the Second World War, Elton used his knowledge of rodents to write reports for the government on improving the use of pesticides to reduce food loss to rats, mice and rabbits – all of them, as he noted, aliens from continental Europe.[2] From this work, he developed a rather doom-laden view about migrating species in general. He went public with these fears, first in three BBC radio programmes broadcast in 1957 and then the following year in a book, *The Ecology of Invasions by Animals and Plants.*

The book began with the observation that 'Nowadays we live in a very explosive world … It is not just nuclear bombs and wars that threaten us … this book is about ecological explosions.' It went on to describe the spread and impact of a range of migrant species, some familiar today like the starling in North America and the Chinese mitten crab in Europe, but others largely forgotten, like the muskrat and the cabbage butterfly. Elton called their spread 'one of the great historical convulsions in the world's flora and fauna'. He used the words of Sir Arthur Conan Doyle in *Lost World* to describe the task of repelling the invaders as 'one of the decisive battles of history'.[3]

Elton was not the first to use such militaristic language about alien species. It had pervaded thinking about nature in Nazi Germany, notes Michael Barbour of UC Davis.[4] A leading German botanist of the day, Reinhold Tüxen, argued that purging the biological invaders would 'cleanse the German landscape of unharmonious foreign substance'. A common Eurasian weed, *Impatiens parviflora*, was condemned as a 'Mongolian invader' that should be repelled 'as with the fight against Bolshevism'. In advocating native plants along the Third Reich's new autobahns, 'Nazi architects explicitly compared their proposed restrictions to Aryan purification of the people', said evolutionary biologist Stephen Jay Gould, whose mother was Jewish. While nobody would accuse today's environmentalists of being secret fascists, this political legacy is, as Barbour put it, disquieting. 'How slippery the slope', said Gould, between love of the familiar and hatred of the foreign.[5]

After the Second World War ended, the Cold War seemed only to increase the cultural hysteria about mysterious and threatening invaders, which often involved alien plants as themes. In John Wyndham's 1951 novel, *The Day of the Triffids*, a South American plant genetically engineered by Russian scientists takes over the world after learning how to walk and kill humans. In the 1956 film *Invasion of the Body Snatchers*, automaton versions of humans hatch from giant seed pods and take on the identity of the first person they encounter.

Despite such cultural resonances, Elton's book did not immediately attract much attention. But it did gain traction as wider environmental concerns grew at the end of the 1960s. Stanford University's Peter Vitousek, who has been a prominent writer on alien species throughout his career, told *New Scientist* magazine that reading *The Ecology of Invasions* as an undergraduate in 1969 'persuaded me not to be a political economist but to be an ecologist'.[6] It inspired others of his generation like Paul Ehrlich, author of *The Population Bomb*, and is often set alongside Rachel Carson's *Silent Spring* as a key scientific text that underpinned new environmental thinking. It also spawned popular books on alien species themselves, such as Carolyn King's *Immigrant Killers*, which described their impact on native birds in New Zealand, and Alexandre Meinesz's *Killer Algae*, on *Caulerpa taxifolia* in the Mediterranean.[7]

Such excited language, replete with military and xenophobic metaphors, has continued to feature in the everyday discourse of scientists investigating alien species, and even in their research papers. It may not sound very dispassionate or scientific, but it reflects the assumption, widespread since Elton, that foreign species are up to no good, and that their alienness means their impacts can be assumed to be bad. They are guilty until proved innocent. In scientific journals where researchers normally strive to use neutral language, those who call themselves 'invasion biologists' stand out. The subjects of their inquiries 'explode' on arrival, killing, eradicating, assaulting and decimating native species, while over-running, flooding and devastating their new habitats. They secure beachheads and fight battles.

Daniel Simberloff of the University of Tennessee, currently the most prominent successor to Elton, began an article for the

National Academy of Sciences on 'biological invasions' by warning in large type that 'An army of invasive plant and animal species is over-running the United States', and noting darkly that the zebra mussel had come 'from the former Soviet Union'.[8] The demise of the Soviet Union did not change the rhetoric much. After 9/11, researchers were soon describing alien species as conducting terrorist attacks on the environment.

Simberloff admits that 'some US nativists in the past lumped introduced species with human immigrants as objects of scorn'. But he insists that 'this does not mean that everyone concerned about introduced species was a xenophobe or racist'.[9] Of course not. But some have been. And it does not help his case that Cornell's David Pimentel, whose influential economic demonizing of alien species a decade ago we explore later in this chapter, was at the same time a prominent supporter of a faction within the conservation group the Sierra Club that opposed any further human migrants from Latin America to the US. With human and biological aliens somehow fused in the popular imagination, it was hardly surprising that George W. Bush, as president after 9/11, moved staff responsible for invasive species from the Animal and Plant Health Inspection Service into his newly created Department of Homeland Security, whose mission is 'to secure the nation from the many threats we face'.

*

In the past quarter-century, 'invasion biology' has become a distinct academic discipline, with its own journals, conferences, research centres and gurus, such as Elton and Simberloff. But there is growing criticism of its narrow agenda and apparent myopic focus on demonstrating the hypothesis that aliens are bad. The charge is that invasion biologists have shown systematic bias in their studies. They have started from the presumption that alien species are bad, and sought out research topics that confirm their view. Like tabloid editors, they concentrate their studies on the nastiest and most sensational invaders. They have rarely been open to other interpretations, and rarely investigate aliens with beneficial or more nuanced impacts on their surroundings.

Research into the researchers seems to back this up. A 2008 study of papers published in a range of invasion biology journals found that 49 species had been the subject or ten or more studies. Almost all were well-known troublemakers, headed by the zebra mussel with 64 papers and the Argentine ant with 61. Other high fliers included the cane toad in Australia, the Mediterranean 'killer algae' *Caulerpa taxifolia*, and Europe's wild boar. It could be that these species are representative of alien invaders as a whole, and that there are hundreds of other unstudied species that are an equal menace. But it is an accepted rule of thumb that at least 90 per cent of invaders quickly disappear and of the remainder only around 10 per cent cause any trouble. So it seems that research is overwhelmingly concentrated on that 1 per cent of the total. The authors of the 2008 findings, including David Richardson of Stellenbosch University in South Africa, were surely right that 'it is the impact of the species that largely determines whether or not it is studied'.[10]

There is similar bias in geographical areas covered. Hundreds of papers explore the impact of invasive species on the handful of tiny islands known to have had their natives most heavily challenged by aliens. Simberloff's 2013 book, *Invasive Species: What Everyone Needs to Know*, has 38 pages on invasions to Hawaii and 36 pages on South Florida – both tiny specks on the planet. Yet only 22 pages include references to the continent of Africa, and most of them are about invasions from Africa to other places or alien species in South Africa, which is one country among 58.[11] Invasion biologists seem to know little, and care even less, about either the silent majority of migrating species who just fit in, or the great many places where there is little pandaemonium.

Perhaps this is not surprising. Researchers are only human, and bad news is always more interesting than good news. I am a journalist; I know all about that. And, says Jennifer Ruesink of the University of Washington in Seattle, the bias towards the troublemaker alien is inevitable 'because that is the hypothesis they test'.[12] But if that is the only hypothesis they test, then it seems a poor academic discipline. And if general conclusions about alien species are drawn from such a biased research base, then we have a problem. The big claims made for the dangers posed by the generality of invaders may be thoroughly unreliable. So may the conclusion that we have a duty to try to prevent them all. It is scientific mythmaking.

We need some new icons to represent benign aliens. One might be the dandelion (*Taraxacum officinale*), which seems always to have been an unremarked presence across Europe and Asia. It will grow just about anywhere there is sunlight. It travelled everywhere with European imperialists, who believed it had medicinal uses. The British took it to New England, the French to Canada and the Spaniards to California and Mexico. It likes disturbed land of the kind humans specialize in creating. Its puffballs of seeds easily blow long distances. It provides nectar for bees. Not even the most determined bio-xenophobe has come up with any nasty habits that make the dandelion a bad neighbour. One website warns that 'you can probably find it on any block in America'. Since it does little discernible harm, that sounds like good news.

Whether alien species are deemed to be bad neighbours seems remarkably dependent on the whims of fashion and scientific inquiry. Ecological disaster stories of the past often rapidly fade. The villains cease to be villains. Back in the 1970s, coco grass, an old-world herbal remedy also known as purple nut sedge (*Cyperus rotundus*), was given top spot among the 'world's worst weeds' in a 600-page inventory put together by LeRoy Holm, then of the University of Wisconsin, Madison. But it does not even merit a place on the list of the top 100 worst invaders compiled by today's generation of invasion biologists, for the IUCN. In fact, only two of Holm's top ten make it into today's top 100: cogon grass and water hyacinth.[13] Similarly, of the seven prime case studies cited by Elton in his 1958 book, only two – starlings and Chinese mitten crabs – make it into today's top 100.

*

So, behind the demonology, what is the truth? One of the most common charges against alien species is that they almost inevitably cause extinctions among their new hosts. They cause a decline in biodiversity. The theory behind this is rooted in ecological thinking about how natural ecosystems are well-oiled and functioning machines in which every native species inhabits a unique 'niche' with a particular job to do, whether as prey or predator, pollinator or processor of waste. The ecosystem is 'saturated' with species, and there is usually little room for interlopers to make much of a contribution. So if a new species takes root, it will usually supplant a native. Expelled from the system, that native may become extinct, at least locally.

A handful of researchers have tried to quantify this extinction threat from aliens, and their findings have been widely quoted. Britain's non-native species secretariat, a government agency charged with addressing the problem of aliens, gave prominence on its website in late 2013 to the 'fact' that invader species 'have contributed to 40 per cent of the animal extinctions that have occurred in the last 400 years'.[14] I was intrigued by the confidence of this assertion. So I went in search of the evidence that backs it up.

The website sources it to the second Global Biodiversity Outlook report, published in 2006 by the secretariat of the UN Convention on Biological Diversity (CBD) and the UN Environment Programme. That document is a bit more nuanced. It says that invasive species are involved in 'nearly 40 per cent of all animal extinctions *for which the cause is known*'. It says nothing about 400 years. But nor does it offer any research to back up its claim. UNEP referred me to the CBD's programme officer Junko Shimura, who in turn said that the source was a 2005 paper from Cornell ecologist David Pimentel. That paper in turn gave as its source a paper by Princeton ecologist David Wilcove published seven years earlier, when Wilcove worked for the Environmental Defense Fund, a US-based NGO.[15]

Wilcove's 1998 paper, in the journal *BioScience*, is a widely quoted text on invasive species. Google Scholar found more than 2,000 citations for it in other papers. But Wilcove was not saying what UNEP, CBD or the British government claimed. He was making a rather different point. He was arguing, on the basis of a review of papers by invasion biologists, that 49 per cent (not 40 per cent) of the extinction 'threat' to 'imperilled' species came in some part from invader species. So he was not talking about actual extinctions, but an extinction 'threat', as judged by invasion biologists. Even more surprising, given the global claims being made by the international agencies, he was reviewing only papers about US endangered species. He made no claims that the conclusion might apply more widely.

Mark Davis, of Macalester College in Minnesota, who has reviewed the research that Wilcove collected for his paper, says the paper is not even truly a reflection of the situation in the US. 'The high ranking of non-native species as an extinction threat was due almost entirely to the inclusion of Hawaii [which] clearly has a dramatically different invasion history' from most of the rest of the country.[16] The extent of this skewing is remarkable. Hawaii makes up 0.3 per cent of the US, yet was responsible for approaching two-fifths of the 'imperilled' birds and plants in Wilcove's data.[17]

So, in effect, Wilcove presented data from Hawaii as stats for the US as a whole, while subsequent users, among them UN agencies, made out that they represented the whole world. Hawaii does not equal the world. The conclusions are nonsense.

This analysis, incidentally, is also the usually-cited source of the oft-repeated claim that invasive species are the world's second biggest threat to global biodiversity, after habitat loss. That claim may or may not be true, but Wilcove's paper certainly does not show it.

Wilcove is not primarily to blame for this. His paper was scrupulous about underlining how little data he had to go on, saying there were 'few quantitative studies of threats to species'. And he made clear that 'the attribution of a specific threat to a species is usually based on the judgment of an expert source'. When I asked him if he agreed with me that the conclusions drawn by others could not be justified by his paper, he replied: 'You are correct; my 1998 paper focused strictly on US endangered species (not extinct species). I don't know the source of the 40 per cent figure for animal extinctions.' All he knew was that it wasn't him. So, one of the most widely quoted claims about the damage done by alien species apparently has no basis in published science.

I wanted to track down another 'fact' – since deleted – on the UK invasive species secretariat's website. Apparently contradicting the previous one, it said that 'invasive non-native species are a known factor in 54 per cent of animal extinctions and the only factor in 20 per cent'.[18] Again the trail led back to a single source, this time a paper published in 2005 in the journal *Trends in Ecology and Evolution* by Miguel Clavero and Emili García-Berthou of the University of Girona in Spain. Titled 'Invasive Species Are a Leading Cause of Animal Extinctions', it had been cited 444 times in other papers, according to Google Scholar. Not bad for a paper that is just four paragraphs long.[19]

This paper did, at least, deal with extinctions rather than threats. The authors got their percentages by accessing the IUCN's database on known species extinctions and checking out the attached notes on what may have caused them. But the authors admit that only a quarter of the 680 listed extinct species had any such notes. The rest were silent on the matter. So it is not true that non-natives were a 'known factor' in 54 per cent of animal extinctions. Nor even of those extinctions that got into the database. It is similarly untrue that science knows that aliens are the 'only factor' in a fifth of animal extinctions. And, given what we know about the biases in how invasion biologists choose their research topics, it seems to me highly likely that the 170 cases where there is a note on the database about likely causes of extinction will be biased towards cases involving aliens.

I thought I might try to check this potential bias by finding out which species the authors had included in their list. The four-paragraph paper offered no clues. And, when I contacted them, the authors said they could not provide a list. They did not have one. Nor could they say in detail how they had done their analysis. So I could not establish how individual extinction cases reached the threshold of a 'known factor' or the 'only factor'. Nobody I could find had replicated this analysis. And since the database had changed since their study, doing so now was probably impossible. That seems, on the face of it, to break two of the cardinal rules of science: that other researchers should be able to try to replicate the findings, and that nothing should be accepted as fact until it has been replicated.

Clavero and García-Berthou presented their four-paragraph paper as a riposte to a previous – and much more substantial – paper in the same journal by Jessica Gurevitch and Dianna Padilla of Stony Brook University in New York. They had argued that 'available data supporting invasion as a cause of extinctions are, in many cases, anecdotal, speculative and based upon limited observation'.[20] Alien species, they pointed out, often turn up in a new place at roughly the same time as natives decline. But that does not prove cause and effect. It is equally likely that some other change – perhaps to climate or the local habitat – messes with the natives, and the aliens just move in to fill the gap. We have seen this process repeatedly in previous chapters. Some researchers say this makes aliens 'passengers rather than drivers'.[21] That makes sense to me.

There is a threadbare laxness in the use of statistics by many invasion biologists. Almost wherever I pursued a key claim, the trail fizzled out in obfuscation, false citations, unverified judgment calls and absurd leaps from the specific to the general and the local to the global. Several trails lead back to genuinely important and pioneering work by Pimentel. He is a revered figure among ecologists, and I have quoted and cited him many times in writing about many scientific topics. He does pioneering work. But here this status seems to have become a liability. He is so highly regarded that nobody tries to replicate his findings or update them. Years later, his back-of-the-envelope calculations are handed down like tablets of stone. That really is not his fault; it is the fault of the wider scientific community.

Just over a decade ago, Pimentel published a series of papers trying to estimate the costs to the US and the world of invasive species. I will return to the economics in the papers later in this chapter, but for now let's stick to his observations about the environmental problems associated with non-native species. In a 2001 paper on the global threat, he stated in conclusion that 'an estimated 20–30 per cent of the introduced species are pests and cause major environmental problems'.[22] This statement has been widely cited by others, including (once again) the UK invasive species secretariat and official documents of the CBD. But the figure did not seem to result from any research by him in the paper, so I asked for his source. His emailed reply said: 'This 20–30% figure is mine, based on my long-time experience.' That's it. We have to take his word for it, because nobody has troubled to try to improve on it.

In a 2004 paper on US alien species, Pimentel noted in passing that 'in other regions of the world, as many as 80 per cent of the endangered species are threatened and at risk due to the pressures of non-native species'. This too is widely quoted by the CBD and others.[23] Here he had a source: 'Armstrong 1995', which turned out to be one sentence in a news item by my old

New Scientist journalist colleague, South Africa correspondent Susan Armstrong.[24] She was referring specifically to plants in the *fynbos* area, a small part of the Western Cape that is of exceptional biodiversity value, but hardly a 'region of the world'.

I checked back. Armstrong told me her source was Brian van Wilgen of the Centre for Invasion Biology in Stellenbosch. Van Wilgen told me: '80 per cent is wrong; it looks like about 25 per cent of threatened species are [on the threatened list] at least in part because of aliens.' A UNEP study quoting a report from South African scientists (among them van Wilgen) says that just 750 of the 8,574 native plant species of the *fynbos* 'are currently facing extinction [as a result of] pressure from invading species'. That's less than 10 per cent.[25] Whatever, Pimentel's stat has entered the domain of constantly recycled 'facts' about alien species.

Despite the unreliable stats, it might still be true that alien species often cause extinctions. David Wilcove told me: 'Given how harmful rats, cats and other non-native species have been to island faunas, I am sure the proportion of species driven to extinction by non-native species is high.' As an adjunct to this, Simberloff argues that under sustained pressure from invaders, even mature ecosystems will eventually suffer 'invasional meltdown'. He says that 'chronic exposure to many invaders ultimately will undermine almost any ecosystem'. But others say the carnage on a handful of islands, assiduously documented by invasion biologists, does not represent the wider picture. It is true that – as when rats meet ground-nesting birds with no defences – alien predators can cause extinctions. But competition between species for jobs within ecosystems rarely results in one species disappearing, even on islands.

In the past decade or so, a new body of evidence has emerged suggesting that some of the most widely used statistics in the canon of invasion biology do not stand up. When Dov Sax of Brown University conducted a global study of species invasions on oceanic islands, he found that while the number of birds did sometimes go down after invasions, plant biodiversity usually rose. Typically such islands now have twice as many plant species as before.[26] That study almost instantly put Sax at the forefront of a quiet revolution in which a new generation of researchers questioned the demonizing of aliens by Simberloff and others.

Even remote 'problem' islands may gain overall biodiversity. In Hawaii, introductions heavily outnumber extinctions and overall biodiversity has risen. In New Zealand, some native species have been ravaged by invaders, but there has been an overall doubling of plant biodiversity – from around 2,000 species to around 4,000 – thanks to the newcomers brought by Europeans. The biodiversity of birds in New Zealand is as high as it has ever been. Yes, there have been some tragic losses. It is often said that sparrows and starlings hardly make up for the loss of Haast's eagle, the world's biggest eagle, or its principal prey, the large flightless moa. Indeed. But both those birds died out around 600 years ago, as a result of hunting by the Maori people rather than species introductions.[27]

Other prominent figures questioning the widely circulated statistics on the biodiversity threat posed by alien species include Mark Davis, who began to question the basis of invasion biology. It was, he said, based on a false proposition. Except in rare cases, aliens bring greater biodiversity, not less.

When the Suez Canal opened in 1869, it allowed tropical species from the waters of the Indian Ocean to move into the Mediterranean. And they did. Yet while 250 species of all kinds established themselves, there has been only one recorded extinction.[28] Similarly, when the Panama Canal joined the Pacific and Atlantic Oceans in 1914, biodiversity increased on both sides. North America has more birds and mammal species than when the Europeans first landed. And the addition of some 4,000 plant species there has added 20 per cent to biodiversity and not, so far as is known, resulted in a single plant species being lost. Likewise, the UK's 2,300 additional species have not directly caused any known local extinctions.

Davis questioned the claim that species-rich and mature ecosystems are 'saturated' with species, with every available niche taken. In fact, rather than being resistant to taking on more species, 'typically, species-rich communities ... accommodate *more* introduced species than species-poor communities', he said.[29] More alien species means more natives, rather than fewer – a rule that seemed to apply from New Zealand to the Appalachians.[30] Aliens may find new jobs to do, or share jobs with natives. But either way, there is no shortage of tasks, and biodiversity usually increases. And rather than suffering 'invasional meltdown' as a result of a constant assault from new arrivals, these augmented ecosystems will in effect become inoculated against any threat from future incomers. The greater variety of species in the system will make it more robust and better able to embrace and make use of any new invaders, an idea termed 'biotic resistance'.

*

In the face of this evidence, Simberloff and others offer a revised argument. Yes, they say, this boost to biodiversity in invaded ecosystems sometimes happens. But this is a temporary phenomenon. There is an 'extinction debt' in many disturbed ecosystems. The arrival of aliens will have set some existing species on a cycle of decline that will ultimately lead them to oblivion. It may take decades for new species to establish themselves, and maybe decades more for natives to succumb. But it will happen. Simberloff points out how there is often a time lag of several decades between the arrival of new species and their break-out. The Brazilian pepper-tree in the Florida Everglades 'remained restricted for a century before rapidly expanding'. Other examples might include the crazy ants of Christmas Island or the Nile perch in Lake Victoria.[31]

Many natives may hang on in the face of invasions, but they

are 'the living dead', says David Richardson. 'Many of the positive impacts attributed to non-natives are likely to be transient ... whereas negative impacts are typically more permanent and often irreversible.'[32] Some modelling studies suggest that 'extinction debts may take hundreds of years to play out', says Benjamin Gilbert of the University of Toronto.[33]

These arguments have not been convincingly refuted. That would be hard. Only time will tell. But there is a flipside to extinction debt, say Sax and Stephen Jackson of the University of Wyoming. They call it 'immigration credit'. Just as extinctions may lag, so may new arrivals. Take the example of the European Alps, where plant diversity has increased on mountain summits during the past century. As temperatures have warmed, species have moved upslope, but incumbent populations have mostly persisted. The newcomers took time to disperse, set up, grow and reproduce. This process lagged behind the warming, but it happened. The colonists to come are as inevitable as any natives that may be lost.[34]

Another side of the immigration credit is the rush of hybridization often unleashed as the genes of the old species intermingle with those of the new. With time, new hybrids and eventually new species will emerge. We will return to these aspects in a later chapter. Both the debt and the credit create uncertainty about the eventual impact of the ecological changes. But to count one without the other makes no sense. It reeks of a bias against aliens.

*

Behind all the arguments about how ecosystems react to invaders, there is a more basic issue. The abiding presumption in invasion biology is that there is something inherently different and bad about aliens. That they are at best misfits and at worst a destructive presence. That, as the environment group WWF puts it on its website, they 'do not belong', are 'unwelcome' and have impacts that 'are immense, insidious, and usually irreversible'. Similarly, natives are good.[35] The sentiment feeds the research agenda, which further strengthens the sentiment. It is the bedrock of policy-making. As Matthew Chew of Arizona State University put it after years trying to break the irrational hatred of tamarisk in the American West, 'nativeness is an organizing principle of numerous studies and findings, and the *sine qua non* invoked by many management policies, plans, and actions to justify intervening on prevailing ecosystem processes'.[36]

The battlefield between those who believe in the dichotomy between aliens and natives, and those who think it is myth-making, fills with anecdotal accounts and one-off case studies. Japanese knotweed versus bracken; Gough Island super-mice versus Joseph Hooker's Ascension Island imports; acacia as water-guzzling menace in South Africa or as a greener of deserts to the north. This doesn't get us far. Where is the truth? Are there any general rules to apply?

A few researchers have attempted meta-studies – sifting through hundreds of research papers to find where the balance of evidence lies. Esteban Paolucci of the University of Windsor, Canada, claimed to refute the friends of alien species with an analysis of more than 1,400 papers published in 2010 and 2011 on the relative impacts of alien and native predators. He found that, taken together, 'alien consumers inflict greater damage on prey populations than do native consumers' – 2.4 times more damage.[37]

But a closer look does not give confidence in the robustness of his findings. For a start, Paolucci had to reject all but 62 of the 1,400 papers that he had found, because they contained no 'quantitative descriptions' of the extent of any impact. And of the tiny minority that remained, few directly compared aliens with natives. The majority were about aliens alone. Given the well-attested penchant of invasion biologists for a bad-news story, Paolucci's conclusion seems to me rather doubtful. The same concern about cherry-picking extends to a similar analysis by

Simberloff of published literature in the US. He found that 'six times more non-native species have been termed invasive than native species'.[38] Termed by whom? And on what basis?

All this feels like self-serving nonsense. Researchers whose working assumption is that alien species do harm review their own literature to discover that more aliens than non-natives are 'termed' harmful. It is hard to disagree with Lawrence Slobodkin, of the State University of New York until his death in 2012, who railed against the 'reification' of untested ideas and unvalidated hypotheses among invasion biologists. He asked: 'What can it mean, 150 years after Darwin, to say that some species or communities are good and some are bad?'[39]

We will return in a later chapter to some of the ecological thinking behind such presumptions. But, for now, surely it is at least as valid, and far more even-handed, to start from the belief that aliens are not in any sense bad, but are essentially indistinguishable from natives. Both groups are capable of causing harm, sometimes dramatically, under some circumstances. Moreover, like natives, aliens can do good. If their alienness can sometimes be problematic, then equally it can sometimes revive flagging ecosystems, creating new space for natives and providing ecosystem services. Without very good grounds for taking a contrary view, it is surely more reasonable to treat species on their merits.

*

A final element in the mythology of the badness of alien species is the economics. Conservationists are increasingly keen to translate their concerns about the fate of nature into terms more amenable to debate in company board rooms and ministerial briefings. They want to put dollar signs on the value of conserving nature and on the dangers of trashing it. This is not an easy task. Every economic assessment of nature rests on numerous contestable assumptions, not least how to deal with the economists' conventions on discounting future loss. I sympathize. But even by the conventional standards of environmental economics, much work on alien species stands out as crude and sometimes plain daft.

I began with another 'fact': the frequently repeated statement that the annual cost of alien species to the global economy is more than $1.4 trillion. This works out at 5 per cent of the entire global economy. UNEP boss Achim Steiner has used the figure a lot.[40] Few question it. But the stat is, to put it mildly, built on rickety foundations. The source trail goes back to a single paper, the work of Cornell ecologist David Pimentel, whose work we touched on earlier in the chapter. His key paper is a 2001 study that extrapolated the $1.4 trillion figure from data gleaned from six countries: the UK, Australia, Brazil, South Africa, India, and – from some of his own more detailed work – the US.[41]

Pimentel's US studies put the total annual cost of aliens to the US economy at $120 billion. The three largest elements were crop weeds at $27 billion, crop plant pathogens at £21.5 billion and rats at $19 billion. The last valued each rat at $15, based on estimates of how much stored grain they eat. Thus the total bill of aliens to the US economy was $420 per person. Taken together, he calculated that the six countries had total annual damage from introduced species of something over $200 per human. Scaled up to the world's population, that delivered a global bill of $1.4 trillion.

That calculation begs some questions about how representative those six countries are. But the more immediate question is whether the national calculations hold water. This was a pioneering first approximation, so it would be unfair to be too picky. But my first concern was simply factual errors about aliens. Some are relatively trivial. 'In Britain', he wrote, 'efforts are being made to eradicate the introduced muskrat and coypu. The populations have been significantly reduced, but the eradication effort has yet to succeed.'[42] For the record, the last coypu in Britain was eradicated in 1989 and the muskrats have been gone since the 1930s.

But some errors have significant implications for Pimentel's trillion-dollar figure. Rats, mostly brown and black rats, are responsible for just over half the environmental losses from alien species in his six countries, at $56 billion a year. Almost half of the rats' bill, $25 billion, comes from India, where local estimates put the rat population at 2.5 billion. Fine so far. And each Indian rat may, as he reckons, eat its way through $10 of grain every year. But why, I wonder, is this marked down to alien species at all? The brown rat (*Rattus norvegicus*) is thought to originate in China, so while it has probably been in India for thousands of years, it might just qualify as an alien. But the black rat (*Rattus rattus*) is widely recognized as originating in India and remains the main rat species found there. It is not alien to India; it has no place in his stats. And yet as the main rat found there it must make up a large part of his $25 billion. I put this to Pimentel by email but he declined to reply.

My second concern was how he did his costing of aliens. His US papers give some useful detail on this. Starlings are assessed on their consumption of grain and cherries, dogs for their livestock kills and medical bills for bitten humans, and domestic cats (*Felis catus*, originally from Africa and domesticated in the eastern Mediterranean about 3,000 years ago) for their bird kills. In what seemed to me a rather arbitrary way – based mainly on the amount of money people spend to go bird-spotting or shooting – he assessed the cost of each bird caught by domestic cats and their feral cousins at $30. Multiply that by the half-billion or so birds estimated to be killed by *Felis catus* each year, and the annual cost of cats comes out at a staggering $17 billion. Well, maybe. But in what sense is this a cost to the US economy? In what sense would each American be $50 or so richer without this cat-on-bird carnage?[43]

Some birds feature as a cost when killed by cats, but also as a cost when they stay alive. In his 2001 study, Pimentel puts the cost to the six nations of three common avian invaders – pigeons, sparrows and starlings – at $2.4 billion. Starlings (*Sturnus vulgaris*) in the US alone cost $800 million. Perhaps they do eat crops worth that amount. But, as Pimentel himself notes in passing, 'these aggressive birds have displaced numerous native birds'. Given that, we must presume that the starlings are often eating crops that native birds would otherwise have eaten. But nowhere does Pimentel ask – still less answer – whether their presence actually increases the crops taken by birds at all.

The basic problem, as I see it, is that Pimentel stacks up every cost he can find on the debit side, but nowhere offsets this with an assessment of any mitigating factors, still less the possible economic benefits of alien species. To be fair, he does not claim to have conducted a rigorous cost-benefit analysis. But surely to leave out one side of the ledger is poor economics. And certainly others who have re-used his statistics are guilty of misrepresenting what they do and don't show.

Let's go back to cats in the US. It is well known that they hunt mice as well as birds. By one estimate they kill ten times more mice than birds.[44] Pimentel acknowledges that mice are 'particularly abundant and destructive' in the US. Yet nowhere does he suggest that cats should get a cash credit for eating mice. Taking a wider view, cats also make their owners healthier. A Canadian study found that claims made to health insurance companies by pet owners were a third less than those made by petless people. And if cat owners did get sick, they left hospital sooner.[45] Singling out the bird-eating skills of the alien cat as the only measure of its economic activity seems perverse.

Most valuable aliens get left out of Pimentel's calculations altogether. There is nothing on the credit side for the European honeybee, which the US government estimates carries out pollinating duties worth $20 billion a year; nor for the food value of introduced fish in rivers, the hunting fees garnered from foreign game, or the soil-binding power of introduced plants. The huge economic value of introduced crops is also ignored. Pimentel does say: 'We recognize that nearly all our crop and livestock species are alien and have proven essential to the viability of our agriculture and economy.' But he continues that 'this does not diminish the enormous negative impacts of other non-indigenous species'.[46] Maybe not, but it does seem odd to include among these negative impacts the billions of dollars spent on pesticides to protect one set of alien species from another – including no less than a billion dollars for controlling weeds on the alien greens and fairways of golf courses. And so on.

It may be unfair to single out Pimentel and his shaky statistics. He had a first stab at an important issue. He should be commended for trying. But that was more than a decade ago. The real scandal is that everyone, from the UN's top environmental scientists down, keeps quoting his figures. Nobody has attempted to replicate or better them. This represents a failure of science. If, as many continue to claim, alien species are the second most important threat to the planet's biodiversity, is it too much to ask that someone else has a go at costing them? Why has no UN agency commissioned a follow-up? Do they just like the existing stats too much? In journalism, cynical hacks sometimes recite the tired dictum that 'some stories are just too good to check'. It is distressing to conclude that the global science community sometimes works by the same standards.

Scientists, starting with Elton, should take considerable responsibility for building public sentiment against anything alien, and for giving respectability to some unpleasantly xenophobic language on the issue from environmental groups, as well as ill-considered legislation designed to keep alien species at bay. Invasion biology has taken as read that alien species are an almost universally bad thing, damaging nature, degrading ecosystems and diminishing biodiversity. It has generated UN treaties, such as the Convention on Biological Diversity, agreed at the Earth Summit in Rio in 1992, which lists the 'prevention, control and eradication' of harmful invasive species as a core target. Partly to pursue that agenda, a United Nations conference in 1996 set up the Global Invasive Species Programme, with a mission to 'conserve biodiversity and sustain human livelihoods by minimizing the spread and impact of invasive alien species'.[47] National governments, including the US and the UK, have widely followed suit, though the global programme was shut in 2011 for lack of funds.

As we have seen in this chapter, there is a push-back among many ecologists against this dogma. They say that the whole idea of invasion biology as a distinct discipline is founded on a false premise and obscures more than it reveals. They would wind it up.[48]

But instead it seems to go from strength to strength. When in 2011 the journal *Science* carried a series of articles quoting ecologists who questioned the pervasive fear of invasive species, a full suite of heavy-hitting conservation leaders wrote in reply. They warned that such talk 'risks trivializing the global action that is needed to address one of the most severe and fastest growing threats to biological diversity'. Among the signatories were the heads of WWF International, the IUCN, Conservation International, Birdlife International and the Wildlife Conservation Society. Of the leading groups, only The Nature Conservancy in the US – perhaps under the influence of its heretical science director Peter Kareiva, of whom more later – ducked out of this political crusade. The conservationists' letter to the world's leading science journal concluded imperiously: 'Now more than ever, academics should be supporting action against invasive species.'[49]

The juggernaut continues. In 2014, the European Union approved new legislation that would ban the possession, transport or sale of 'alien invasive species' that officials deem to be of 'Union concern'. As of late 2014, there was no list of these species, but the phrase 'alien invasive species' is gaining favour. It is a catch-all for nastiness, and a recipe for muddled thinking. Alienness is about where a species comes from or 'belongs'; invasiveness is about its behaviour. Conflating the two feeds the idea that all aliens are invaders, and all invaders are aliens. There is, as we have seen, little basis for either assertion.[50]

8
Myths of the Pristine

Take a journey up the River Amazon and you are soon floating past river banks festooned with primeval rainforest unchanged for millions of years. So say guide books and many conservationists. Except that you aren't. There is jungle, for sure. Despite the large areas of forest that have been cut down in recent decades, most of it remains. But little of it is primeval. Certainly not along the river banks which, until European explorers first came here 500 years ago, were centres of trade, agriculture and urban settlement. In thinking otherwise, we have succumbed to the myth of the pristine.

When the Spanish conquistador Francisco de Orellana sailed the Amazon in 1542, his chronicler, Friar Gaspar de Carvajal, wrote in his journal of a place some 1,000 kilometres upstream, near the junction with the Rio Negro. 'There was one town that stretched for fifteen miles without any space from house to house, which was a marvellous thing to behold', he wrote. 'There were many roads here that entered into the interior of the land, very fine highways. Inland from the river to a distance of six miles more or less, there could be seen some very large cities that glistened in white and besides this, the land is as fertile and as normal in appearance as our Spain.'[1]

That city was never seen again by outsiders. Later explorers reported finding only small bands of Indians hiding deep in the jungle. But Carvajal's journal was no jungle dream. In recent years, archaeologists have discovered hundreds of large earthworks in the Amazon rainforest that they conclude were urban centres that flourished until the conquistadors came, bringing European diseases and European rule. Anna Roosevelt, an archaeologist at the Field Museum of Natural History in Chicago, called them 'one of the outstanding indigenous cultural achievements of the New World'.[2] Michael Heckenberger of the University of Florida said they showed the presence of a 'highly elaborate built environment, rivalling that of many contemporary complex societies'. Far beyond the city limits, the jungle had been transformed. Most of this 'primeval' forest had been cleared at least once, and perhaps several times, for farming. It was a living landscape. What is today one of the largest tracts of rainforest in the world was, until less than 500 years ago, a chunk of tropical suburbia.[3]

It is now estimated that before 1492, there were more than 50 million people in the Americas, half of them in South America. They were spread widely and used the land extensively. They burned vegetation to encircle animals for hunting, and to increase the productivity of grasslands. And they certainly didn't spare the rainforests. Clark Erickson of the University of Pennsylvania has found tens of thousands of kilometres of raised banks across the Bolivian Amazon that he believes were dug by farmers to keep their maize, manioc and squash crops clear of seasonal floodwaters and highland frosts. They were interspersed with causeways and islands that may have been human settlements, and certainly contain many human remains.[4]

Erickson's findings seem to corroborate the record of a failed Spanish expedition to conquer Baures in the lowlands of Bolivia in 1617, which described entering towns along causeways that could take four horse riders abreast. Jesuit records suggest that some islands and causeways remained in use into the 18th century, before being abandoned to regrowing forest and populations of tapirs, peccaries and deer. Similar remains turn up from Peru to the Guyanas. The earth-moving involved in creating them must have been 'comparable to building the pyramids. They completely altered the landscape', Erickson told me. 'Some people want to preserve the forests. That is fine by me, but there is no way they are pristine. Every feature of this land is man-made.'

*

The idea of the tropics largely comprising a pristine wilderness seems to go back to the 19th century romantic writers and painters. For a long time, scientists bought into it too. But recently, from the Americas through Africa to South-east Asia, forest researchers have been finding extensive evidence of past forest clearance, for cultivation and even to fuel industrial activity such as metal smelting. The rainforest civilizations of the Mayans in Mexico and Angkor Wat in Cambodia are famous. But there is much more. Archaeologists are finding the remains of complex urban civilizations in apparently pristine West African forest. It also seems that, until 1,500 years ago, much of the Congo was cleared of jungle.

Researchers are coming to realize that humans have sculpted the deepest jungle for thousands of years. Many rainforests – perhaps all – are partly regrowth and partly a result of deliberate planting. Rather than wilderness, they are abandoned gardens. The widely held notion that civilization got going only in drier climates – and that rainforest dwellers were happy to live a semi-nomadic life of hunting and gathering – is being kicked into some very long grass. But the reappraisal has been hard-fought. The idea of pristine nature is deeply ingrained among conservationists and many academics.

Witness the new story being pieced together about central Africa. Mike Fay, a young American explorer who has survived a plane crash, confrontations with armed rebels and being gored by elephants, is one of the few people in modern times to walk – and wade – right across the swampy rainforests of the Congo basin in central Africa. The 2,000-kilometre trek took him fourteen months. Afterwards he told me: 'I have never come across a virgin forest in Africa. It is obvious that man has been a player for a very long time.' The forest floor was almost everywhere littered with oil palm nuts, he said. They were the remains of plantations subsequently dated to some 2,000 years ago. In the Central African Republic he found layers of charcoal under forest soils, as well as pottery fragments and the remains of an iron smelting industry that used trees as fuel eight centuries ago.

When Fay later flew his Cessna plane over the forest, he kept noticing small grassy mounds that he is convinced are the remains of ancient farming systems.[5] French researcher Germain Bayon has found sediments at the mouths of Africa's great rivers that reveal widespread deforestation across central Africa between 2,000 and 3,000 years ago. This coincided with the migration of Bantu-speaking farmers out of present-day Nigeria and across central Africa, where it seems that they pushed aside nomadic hunter-gatherers and chopped down trees to grow millet, yams and oil palm.[6]

Another French archaeologist, Richard Oslisly of the Institute of Research for Development in Marseille, has found hilltop iron smelting towns in the middle of modern-day Gabon that were open for business more than 2,000 years ago.[7] 'It seems that extensive archaeological remains are hidden beneath much of the African rainforest, suggesting that human disturbance has been one of the dominant factors affecting forest structure and composition in recent millennia', says John Oates of Hunter College, New York.[8] Environmental historian Kathy Willis, now director of science at London's Kew Gardens, takes a similar view. 'Much of the Congo basin underwent extensive habitation, clearance and cultivation beginning around 3,000 years ago and ending around 1,600 years ago', she says. Then something happened. There was certainly a population crash. Perhaps these industrial cultures ran out of trees. Perhaps climate changed. But whatever happened, the jungle closed in.[9]

This may have happened more than once. Barend van Gemerden of Wageningen University in the Netherlands found that in the rainforests of southern Cameroon, almost all the trees that are more than 300 years old are of species that today grow only in clearings made by shifting cultivators. But younger trees are of species that prefer closed forest. His conclusion is that the forests were once full of farmers cutting and burning and cultivating on a large scale. The end came at around the time that Europeans first showed up looking for slaves.[10]

*

What of the other jungle regions of the world? We know that the Angkor civilization ruled much of forested South-east Asia from the 9th to 15th centuries. The extraordinary Angkor Wat temple complex on the shores of Tonle Sap lake in Cambodia was the ceremonial heartland of a densely populated area with networks of roads, canals, reservoirs and rice paddy fields that covered at least a thousand square kilometres. The entire urban jungle must have housed hundreds of thousands of people. There were miners and weavers, boat builders and blacksmiths, traders and priests.

Angkor was likely the pinnacle of a series of early urban societies in that region. The evidence of widespread forest clearance and planting goes back thousands of years, says Chris Hunt of Queen's University Belfast. When he analysed traces of pollen left in the forests from Thailand and Vietnam to the Indonesian islands of Borneo and Sumatra, he found 'a pattern of repeated disturbance of vegetation since the end of the last ice age'. In the highlands of Borneo they had been burning the forest and planting fruit trees. On the coast, he found pollen from the sago palm shipped there from New Guinea 2,000 kilometres away and planted about 10,000 years ago. (He noted angrily that laws in some South-east Asian countries do not recognize the rights of indigenous groups, such as the Penan in Borneo, because they leave no permanent mark on the landscape. Yet they have been leaving their mark for 10,000 years.)[11]

Cut to Central America, where the most famous jungle remains come from the ancient Mayan civilization. Giant pyramids loom out of the forest as a reminder of a culture that occupied large areas of Guatemala, Honduras and southern Mexico from some 3,000 years ago. Population densities in the area then were greater than today. The Mayans planted their forests with *Manilkara* trees, which produce a latex that later became the basis for chewing gum, and millions of giant *Brosimum alicastrum*. This relative of the mulberry tree yields nuts, now known as Maya nuts or breadnuts, which formed a large part of the Mayan diet. The trees remain. But the descendants of the Mayans have forgotten about this forest food so completely that they trample the fallen nuts underfoot and chop down the trees to make room for fields of corn that produce far less food.

In 2006, I helped judge an environmental award given to Erika Vohman, an American who had rediscovered the true worth of Maya nuts while working as a biologist in Guatemala. 'It's the largest tree in the rainforest – up to 45 metres tall', she said. 'It is hard for some people to imagine that a large tree growing in the wild is cultivated. But that is what it is: an ancient cultivated crop.' Thanks to her efforts, thousands of village women are once more harvesting the nuts, making soup, ice cream and cookies for their families, and selling them in local markets. A campaign is under way to save the Maya nut forests.[12]

Once seen as a mysterious aberration, it now seems as if the Mayans, like the empire builders of Angkor, were far from alone in the jungle. Deborah and David Clark were for many years the joint directors of the La Selva forest research station in the Central American state of Costa Rica. US scientists had been working there for years to describe its unchanging ecosystems and burgeoning biodiversity when, twenty years ago, the Clarks emerged to declare this idea a myth. Buried in the soils in the heart of the forest, they had found charcoal, corn pollen and farm tools. The whole area had been used for agriculture for most of the past 2,000 years. The pristine forest was little more than a garden gone wild, abandoned when Europeans first invaded Central America in the wake of Columbus. 'There is no such thing as virginity out there', Deborah Clark, now at the University of Missouri-St Louis, told the *New York Times*.[13]

One reason modern scientists have been reluctant to believe that great civilizations lived in the rainforest is rainforest soils, which their studies show to be generally too poor to sustain large populations. But that has changed with the discovery in the Amazon of extensive patches of what archaeologists have dubbed 'dark earths', or *terra preta*. They appear to be man-made, the product of mixing the natural thin, acidic rainforest soils with a mulch of organic waste and partly burned plant material, rather like charcoal. Both fertilize the soil. The mulch has enriched the soils with micro-organisms that ensure that it regenerates itself even as it is used for cultivation. 'It's at least as good as manure', says Bruno Glaser of the University of Bayreuth in Germany.

The dark earths are often full of pottery shards and other detritus of civilization dating back as much as 2,500 years. Many forests that grow in these dark earths are full of Brazil nuts, lianas, palms and bamboo, as well as fruit and medicinal and other economically valuable forest species. These are thought to have been deliberately planted in times past. All this suggests an unexpected truth, says Susanna Hecht of UC Los Angeles. 'People have been farming there – farming hard – for thousands of years.'[14]

Now that scholars know where to look for dark earths, they are turning up everywhere. British anthropologist James Fraser of the University of Lancaster began his career finding them in the Amazon and is now doing the same in West Africa.[15] In just four months, he found patches at 150 sites across the north of Liberia. In Wenwuta village in Lofa county, 'you come down a path and over a stream, and immediately you come to a ring of cocoa and cotton trees. You can see the dark earth.' The villagers know about the soils, he says, and actively use them. 'If you look closely in the journals of Victorian explorers and agronomists in the 19th and 20th centuries, you find occasional reports of farmers burying wood and other vegetation, covering it with soil and leaving it to smoulder before distributing the resulting ash across their fields. But nobody looked for a pattern or wondered about their significance.'

In 2012, dark earths were located in northern Borneo, on riversides where humans congregated. 'Local people value these soils for cultivation but are unaware of their origins', said study author Douglas Sheil, a tropical forest researcher now of the Norwegian University of Life Sciences.[16] They also turn up outside forests, especially in wetlands. Paleoecologist Gail Chmura of McGill University, Canada, told the American Geophysical Union in 2011 that garbage mounds left by prehistoric humans might have formed the tree islands of the Florida Everglades. They are known for their exceptional species richness and were presumed natural. But they are full of bones, discarded food, charcoal and bits of clay pots. 'This goes to show that human disturbance in the environment doesn't always have negative consequences', says Chmura.[17]

*

The conventional view has for a long time been that the planet – and especially the tropics – was largely unaltered by humans until recently. But that looks increasingly wide of the mark. Most of the world is profoundly altered by human activity often stretching back thousands of years. We knew what we were doing, says Erle Ellis, a geographer at the University of Maryland, Baltimore County. From the tropical rainforest to the tundra, from the tussock grasses of New Zealand to the once-forested Scottish peat bogs, from the floodplains of China to the heaths of central Europe, we have been transforming landscapes, wetting and drying, foresting and deforesting, planting and burning, grazing and ploughing, hunting megafauna to extinction and transporting new species in their place.

Our early hunting exploits were probably responsible for major extinctions among the planet's largest creatures, the sabre-toothed tigers and woolly mammoths, the two-tonne sloths and mastodons. Faced with a decline in big game between 30,000 and 20,000 years ago, the tribes of Europe and Asia started to diversify their food sources by hunting a wider range of smaller animals. They started to manage wild herds of deer and gazelle to ensure they did not die out. They extended their diet further by learning to grind, boil, ferment and roast food. And they developed early techniques of farming, such as seed propagation. It was an early example of adopting a more sustainable lifestyle to cope with resource shortages.

British author Colin Tudge argues that by then, humans were managing the land enough to be called 'proto-farmers'.[18] They were planting primitive crops and raising domesticated animals even as the last ice age ended and ice sheets retreated across North America, northern Europe and Asia. They were cultivating squash by 10,000 years ago, maize by 9,000 years ago, and by 8,000 years ago some European farmers were using manure from sheep and goats to fertilize crops of wheat and barley, and to maintain soils.[19] We were consciously engineering ecosystems, using fire to encourage annual plants that attracted game, for instance.

The bison-grazed plains of North America were remade by Native Americans setting fires long before Europeans showed up. And the Australian outback was similarly treated by Aborigines using their fire-stick farming systems.[20] The mist-shrouded and treeless grasslands of the tropical Andes, known as the *páramos*, are also thought to be the result of burning and grazing after locals cut down the natural forests centuries ago. In the Scottish highlands, the moorland wilderness loved by natives and visitors alike is a result of ancient deforestation and recent depopulation through the notorious 'highland clearances', forced displacements during the 18th and 19th centuries. Erle Ellis calculates that at least a fifth of the Earth's land surface was transformed by humans as early as 5,000 years ago.[21]

But farming was primitive and soils were often quickly exhausted. So we humans used a lot of land even though there were relatively few of us – tens of millions at most, says Steve Vavrus of the University of Wisconsin-Madison. Our ancestors spread out, leaving the cleared land fallow for twenty years or more before returning to take another crop. As we have seen, they farmed large areas that today look like virgin forests.

'Conservationists in many parts of the world are attempting to conserve what they take to be wilderness, but increasing evidence suggests that the territories involved are or were cultural landscapes', created as much by human activity as by nature, says Cambridge botanist David Briggs.[22] His Cambridge colleague, the British landscape historian Oliver Rackham, agrees: 'Most of the world's land surface results from long and complex interactions between human activities and natural processes.'[23]

We live in what geologists are now calling the Anthropocene, an epoch in which the planet is shaped primarily by mankind. We need to get used to it. But we shouldn't be depressed. For while one lesson of the Anthropocene is that nothing is pristine, another is that nature is resilient and resourceful. And while many endangered species are vulnerable to our activities, others really rather like us. Mankind is not always bad news for nature. Some forests died, but others grew up in their places. By breeding some species as crops and livestock, we increased their genetic diversity. And by moving species around the world we dramatically increased local biodiversity in many places, and may sometimes have triggered a burst of evolution. The discovery of the longevity and extent of our influence underlines the resilience of nature to our depredations.

We have assaulted forests on a big scale. Yet where we have

walked away, they have generally revived. That is especially true in the tropics, the area of our greatest current environmental concern. 'So-called virgin forests have in fact undergone substantial prehistoric modification', says Kathy Willis. 'Tropical forest ecosystems are not as fragile as often portrayed, and in fact are quite resilient. Left for long enough, forests will almost certainly regenerate.' There is, she says, no reason why that should not remain the case in the 21st century.[24] The new forests won't be pristine. But then, they never were.

*

I end this chapter with a double-edged story. It shows the dramatic scale of the changes that humans can cause to the environment by an apparently unimportant introduction. But it also shows the extent to which we can be misled about what is natural and what is not. The man who encapsulates that misreading of the land is Teddy Roosevelt, former US president and supreme outdoorsman, in a great tradition of men who liked both to conserve nature and hunt it.

In 1909, Roosevelt went on an infamous safari. He, his gun-toting son Kermit and a rogue's gallery of British imperial hunters set out with more than 200 porters from the port of Mombasa in Kenya. Their year-long spree in the African bush involved daily killing; they eventually shipped more than 11,000 carcases back to the Smithsonian Museum in Washington, DC. Roosevelt observed with sometimes touching sympathy the wild animals he met, before stalking and killing them. This 'great adventure' in what Roosevelt called 'the greatest of the world's great hunting grounds' was big news back home. It cemented both his reputation as macho adventurer as well as politician, and our own image of Africa as somewhere primeval and apart from the rest of the world.[25]

Roosevelt called what he travelled through a Pleistocene landscape. Most of us still have the same vision of a wild African bush teeming with wildebeest and elephants, lions and zebras. We've seen it on TV in endless wildlife documentaries. We often mourn its retreat, first at the hands of European hunters and later as Africa's indigenous population has soared. But, strange to say, like our presumptions about pristine rainforest, it is a myth. Pre-European Africa wasn't like that. Africa's jungles and savanna grasslands – the surviving fragments most redolent of what we imagine to be pristine Africa – are almost as artificial as an English country garden. Africa's gardener, the force that shaped today's landscape, was a microbe from Asia that arrived little more than twenty years before Roosevelt. That day changed the continent for ever.

In 1887, a small force of Italian soldiers made a foray into the Horn of Africa during Europe's colonial-era 'scramble for Africa'. They didn't get far. But the cattle that accompanied them as a mobile source of food carried a vicious hitchhiker – the virus that causes a disease called rinderpest, which is German for 'cattle plague'. The virus is native to the steppes of central Asia and periodically swept through Europe in the Middle Ages, usually with cattle feeding armies during military campaigns. But it was unknown in Africa until it showed up on the Red Sea coast of modern-day Eritrea, at the site of the Italian invasion. It spread through the animal herds of Tigre province in Ethiopia in 1888, and then followed ox trails south down the Rift Valley and west along herding routes through Sudan and Chad into West Africa. Cattle in their millions suffered fever, diarrhoea, lymphoid necrosis and death. Within five years, the virus had reached the shores of the Atlantic. The British colonial authorities tried to halt its advance south by erecting a 1,500-kilometre fence and shooting cattle. But within a decade, it was in South Africa, and its conquest of the continent was complete.[26]

The virus spread to many cloven-hoofed wild animals, including buffalo, wildebeest, eland, giraffe and many species of antelope. The pandemic has been called the greatest natural calamity to befall Africa. 'Never before in the memory of man, or by the voice of tradition, have the cattle died in such vast numbers; never before has the wild game suffered', wrote Frederick Lugard, a British army captain who travelled the caravan routes of northern Kenya in 1890 while scouting for an imperial invasion of East Africa. 'The enormous extent of the devastation can hardly be exaggerated.'[27]

As the cattle died, so did humans. Herders had no livestock and farmers had no oxen to pull their ploughs or drive the waterwheels that irrigated their fields. Hungry people fell prey to native diseases such as smallpox, cholera and typhoid. 'Everywhere the people I saw were gaunt and half-starved, and covered with skin diseases', wrote Lugard. 'They had no crops of any sort to replace the milk and meat which formed their natural diet.'[28] In places, the epidemic coincided with drought. Between 1888 and 1892, roughly a third of the population of Ethiopia – several million people – perished. In Kenya the Maasai tribe still tell of the *enkidaaroto*, the 'destruction'. Most of their cattle died, along with the buffalo and wildebeest of the Serengeti. They fought wars over the surviving cattle and were reduced to begging for meat from passing caravans. One elder later recalled that the corpses of cattle and people were 'so many and so close together that the vultures had forgotten how to fly'.[29] An Austrian explorer, Oskar Baumann, who travelled through Tanzania in 1891, estimated that two-thirds of the Maasai died.

Until the arrival of rinderpest, Africa's richest and most aristocratic kingdoms were built on cattle. But in that decade of carnage, the great herds were wiped out. Rinderpest depopulated and impoverished Africa on a scale that even the slave traders had not managed. Famine decimated cattle-herding tribal kingdoms like the Tutsi in Rwanda, the Soga in Uganda and the Nama and Herero in south-west Africa. In northern Nigeria the Fulani, 'having lost all, or nearly all their cattle, became demented', according to a contemporary account collected by researcher

John Ford. 'Many are said to have done away with themselves. Some roamed the bush calling imaginary cattle.' Much the same befell the Dinka and Bari in Sudan and the Karamajong of Uganda. Many of these societies have never recovered. They never had a chance. For the disease served up the continent on a plate for colonial takeover. In its wake, Europe's 'scramble for Africa', which had been under way since the 1880s, reached its victorious climax.[30]

With warrior tribes barely able to put up a fight, the Germans and British finally secured control of Tanzania and Kenya. Lugard noted blandly that rinderpest 'in some respects has favoured our enterprise. Powerful and warlike as the pastoral tribes are, their pride has been humbled and our progress facilitated by this awful visitation. The advent of the white man had not else been so peaceful.'[31] In southern Africa, the Herero succumbed to German occupiers, and the hungry and destitute Zulus were forced to migrate to the gold mines of Witwatersrand, helping to create the brutal social divide between black and white from which apartheid sprang.

The ramifications of the rinderpest epidemic did not stop with European takeover. For the 'cattle plague' also opened the way for an invasion by an indigenous pest. The tsetse fly (*Glossina palpalis*) has always lived among wild animals in Africa's lowland tropical bush. It carries *trypanosomiasis*, a parasitic disease that, like rinderpest, is lethal to cloven-hoofed animals. Unlike rinderpest, it also attacks humans, in whom it is called sleeping sickness, a disease that causes fever and, without treatment, organ failure and death. Before the arrival of rinderpest in Africa, the tsetse fly and the disease it brought were confined to small areas of the continent, mostly in the Congo rainforest. Tsetse flies loved cattle, but also needed lush vegetation to deposit their larvae. But for centuries, the continent's huge cattle herds had kept the fly in check by grazing the grass close and eating tree seedlings. Rinderpest changed that.

Initially, the rinderpest epidemic was bad news for tsetse flies, because it killed their animal hosts. But without cattle and other grazing animals, the bush grew fast. Within a season or two, the pastures were transformed into woodland and thickets of thorns, where the tsetse could flourish. Meanwhile, wild animal populations rebounded from rinderpest much faster than cattle, so the flies quickly had an animal host to feed on. The landscape was suddenly ideal for the flies, and they spread fast. So did *trypanosomiasis* and sleeping sickness. Cattle could not return to graze down the bush. Huge areas of Africa's former cattle pastures – from the highlands of East Africa to the Zambezi and Limpopo river valleys in the south – quickly became tsetse-infested bush.

The economic damage of this transformation was huge. While rinderpest passed through and largely disappeared, the tsetse takeover prevented millions of cattle-herding people from recovering the wealth they had lost during the epidemic. The impact on human health was even worse. Sleeping sickness was unknown in East Africa before the rinderpest epidemic. But in the aftermath, an estimated 4 million people died in Uganda alone. Sleeping sickness has persisted ever since on a scale not seen before, with some 10,000 deaths a year. And the threat it poses to cattle means that some of the most fertile parts of Africa are still virtual no-go areas for humans and cattle alike. No wonder some conservationists call the tsetse fly 'the best game warden in Africa'.

An alien virus, introduced by humans, did all this. But here is the final point. The rinderpest virus, and the tsetse takeover it triggered, led to an ecological revolution against people and cattle and in favour of wildlife across Africa. It created the bush landscape where Roosevelt conducted his 'great adventure' – a landscape from which cattle had recently been banished and where vegetation and wild animals were newly resurgent. It became the incomers' archetype of 'unspoilt' Africa. But far from being a pristine landscape, it was a recent product of the invasion of a virus brought by humans.

It is no accident that the idea of turning over large areas of empty Africa into game reserves, first for hunting and later for conservation, took hold in the aftermath of the rinderpest disaster. Roosevelt, who had been a pioneer of creating national parks in the US during his time as president, headed a movement to do the same in Africa. Another key figure in the move to conserve Africa's supposedly Pleistocene landscape was Julian Huxley, founding head of UNESCO in the late 1940s and later founder of the World Wildlife Fund. He described the East African plains as 'a surviving sector of the rich natural world as it was before the rise of modern man'.[32]

This new generation of conservationists persuaded colonial authorities to create Africa's great national parks and reserves – the Serengeti and Maasai Mara, Tsavo and Selous, Kafue, Okavango, Kruger and others – from which, they decreed, humans and their cattle had to be excluded. The German biologist Bernhard Grzimek – who produced the famous book and film, *Serengeti Shall Not Die* – worked indefatigably to expel the Maasai from the Serengeti. 'A National Park must remain a piece of primordial wilderness to be effective', he wrote. 'No men, not even native ones, should live inside its borders.' The Serengeti 'cannot support wild animals and domestic cattle at the same time', even though it had long done just that. These eco-pioneers did not realize, or did not care, that half a century earlier, many of these 'pristine' parks had been open cattle pasture.[33]

Today, rinderpest has been exterminated from the planet. But in much of Africa, the world it created persists. There are two ecosystems, created and separated by people. In one, farmers and cattle herders reign and the bush and the tsetse fly are tamed. In the other, the West's vision of 'primeval' Africa holds sway, the bush runs wild and the tsetse flourish. But what is apparently most wild and primeval is actually most recent.

All this tells us three things. First, of course, it tells us about the danger of alien diseases in communities with no immunity – though I doubt that anything on a remotely similar scale still awaits unsuspecting ecosystems today. Second, it tells us that nature is good at growing back. And third, it shows that our attitudes about pristine nature are often founded on false perceptions. Most of what we regard today as virgin wilderness is far from untouched by humans. That should give us pause. If nothing is wilderness, then what does it mean to be a conservationist? What, and how, do we conserve?

Forest ecologist Douglas Sheil says the recognition of this fact should have a profound effect. Conservation scientists are mostly blind to nature outside of what they think of as pristine habitats and routinely ignore its value. To take one emotive example, three-quarters of the world's surviving orang-utans live in timber plantations rather than old-growth forests, and yet 'there are no conservation programmes in active concessions'. Sheil says: 'Conservationists simply cannot bring themselves to accept [that] many species are more disturbance tolerant than is widely assumed.' This blindness, he says, is caused by our yearning for a vision of unblemished nature – a vision that is at odds with reality. 'We should not dismiss the value of modified ecosystems simply because we do not like them.'[34]

9

Nativism in the Garden of Eden

George Perkins Marsh was America's first conservationist. It is amazing that he found the time. He was a polymath from the woods of Vermont, who at different times during the middle of the 19th century practised law, edited newspapers, farmed sheep, cut and sold lumber, lectured students, ran a wool mill, quarried marble, learned twenty languages, designed the Washington Monument, helped found the Smithsonian Institution, served as a Congressman, provided aid for refugees from revolutionary Turkey, and served as US ambassador in Rome. It was while in Rome that he began to ponder how landscapes around the Mediterranean and Middle East had been denuded over the centuries. He concluded that past people had deforested the land, thereby eroding soils, spreading deserts and causing the demise of their own civilizations. And having participated in much the same activity in New England, he feared the same catastrophe could befall the United States too.[1]

So in 1864 Marsh wrote *Man and Nature*, the first manifesto of modern environmentalism. His book preceded more famous texts from later ecological heroes like John Muir and Aldo Leopold. It was a powerful evocation of man as the destroyer of nature's balance. 'Nature, left undisturbed, so fashions her territory as to give it almost unchanging permanence of form, outline and proportion', he wrote. If disturbed 'she lets herself at once to repair the superficial damage, and to restore, as nearly as practicable, the former aspect of her dominion'. But any more fundamental damage was beyond nature's reach. Mankind was the 'disturber of natural harmonies', capable of wreaking change that was irreversible except on the longest timescales. Thinking no doubt of the desertified lands of the Mediterranean, he said that in places man has 'brought the face of the earth to a desolation almost as complete as that of the moon'. Natural recovery, if it happened at all, would take centuries. 'Man, who even now finds scarce breathing room on this vast globe, cannot … wait for the slow action of [nature] to replace, by a new creation, the Eden he has wasted.'[2]

The idea of nature's balance, and how humans are capable of destroying it, did not start with Marsh, of course. It is implicit in the biblical story of the Garden of Eden, in which sinning humans are cast out from the garden. They are separated from nature and doomed to damage her. The idea has persisted right into the modern world in which our view of nature is, ostensibly at least, based on science. The balance of nature has become a 'foundational metaphor in ecology', says Stephen Trudgill, a British geographer from Cambridge University who specializes in what he calls our social engagement with nature. Ecology, he says, is a science built on 'the guilt-laden notion that we have disturbed the natural order, and it is now all wrong and our fault'.[3]

But are we right? Daniel Botkin, an ecologist at the University of California, Santa Barbara, scathingly says that 'overwhelmingly we still believe in the nature as described by the Ancient Greeks, which has come down to us through Judaism and Christianity'. He accuses environmental scientists of failing to question these hand-me-down premises of natural order. He says there is no such thing as a balance of nature. After many years of monitoring local ecosystems like Isle Royale in Lake Superior, he concluded in his book *Discordant Harmonies*, and later in *The Moon in the Nautilus Shell*, that nature is constantly out of balance, constantly changing.[4]

This argument is central to our story about alien species. If nature is, and should be, in balance, then there is no place for alien species. The carpetbaggers from different ecosystems should be sent home. But if there is no balance, no natural order – and no original sin in disturbing nature – then aliens may have a place in the scheme of things. They are not, of themselves, bad things. They may be good. They may deliver. They may, to borrow Darwin's ideas of natural selection, be worthy victors in the survival of the fittest. Much of the rest of this book will explore this idea, what it means for alien species, and what it should mean for the conduct of conservation. Botkin called it the new ecology; I call it the new wild.

*

But first, how did the notion of balance in nature take such a hold among ecologists – and why does it continue to spellbind most in the conservation community? Marsh didn't talk about ecology or ecosystems, but his ideas about nature's propensity for 'unchanging permanence' encouraged his successors to look for the ecological processes that delivered this state of balance. The most influential was the Nebraska botanist Frederic Clements. After years exploring nature in his prairie home state in the late 19th century, the tall, upright and devoutly religious Clements concluded that balance arose from plant species forming tight 'associations' that we would today call ecosystems. For every type of climate in the world, he said, there was a matching permanent association of vegetation. He called it a 'climax'. If something intervened to disturb the balance of that association – whether hurricane or ice age or clear-felling foresters – the association would attempt, through a process he called 'succession', to recreate its former steady state.[5]

On the face of it, Clements's picture of nature as a finely-tuned machine fits Charles Darwin's insights on natural selection. Darwin had proposed in *On the Origin of Species* that many species evolved in harmony to maximize their survival chances. He called it co-evolution. Insects specialized to live off and pollinate plants, while the plants evolved to help them, to ensure their reproduction. Most famously, having been sent an orchid from Madagascar, *Angraecum sesquipedale*, with nectar hidden down a 30-centimetre-long narrow tube, Darwin predicted that there must be an insect somewhere with a proboscis long enough to reach the nectar. Half a century later, just such an insect, *Xanthopan morganii praedicta*, was found. Much later, it was shown to feed on the orchid.[6]

From the idea of co-evolution, it was but one step to suggest that evolutionary forces would act as some kind of optimizing force, creating entire ecosystems in which every species had its tightly defined role. Clements and the early ecologists argued just that. They saw natural selection as creating an evolutionary path to perfected nature. Darwin himself had never suggested any such thing. For him, 'survival of the fittest' was just that. Species evolved and adapted to meet their own needs and gain ascendancy over rivals. But the idea had been proposed by one of Darwin's predecessors in evolutionary theory. The late 18th-century French naturalist Jean-Baptiste Lamarck saw the changes induced by evolution as 'the accomplishment of an immanent [inherent] purpose to perfect the creation'. Clements was a great fan of Lamarck.[7]

Through the early 20th century, Clements' idea of associations of plants was refined into the concept of ecosystems – communities of organisms in which, in ways that have proved hard to define, the whole was greater than the sum of the parts. Ecosystems were 'super-organisms'. The natural world had a metabolism. Here is where our idea of rainforests as systems rather than just a collection of trees comes from – and on a global scale, of a 'biosphere' comprising conjoined ecosystems. The British independent scientist James Lovelock went even further in the 1970s and proposed that the biosphere maintained the planet's atmosphere and oceans in a way fit for life to thrive. Neatly fulfilling Botkin's observation that all these ideas go back to Classical times, Lovelock called this super-organism Gaia, after the Greek earth goddess.[8]

*

This all-embracing ecological orthodoxy has long been attacked by dissident scientists who see it as romantic or even religious flummery with no scientific basis. In his time, Clements' great critic was Henry Gleason. A contemporary of Clements, Gleason also grew up in the Midwest collecting wild plants, but drew very different conclusions. Nature was not composed of climatically determined associations of species, he said. Nature was primarily 'individualistic'. Species did their own thing, according to their genes and physiology, unconstrained by any higher purpose or super-organism straitjacket. 'Every species of plant is a law unto itself, the distribution of which in space depends on its individual peculiarities of migration and environmental requirements', Gleason wrote in 1926. Far from forming discrete and exclusive associations, species 'grow in company with any other species of similar environmental requirements, irrespective of their normal associational affiliations'. There were no exclusive clubs of species maintaining a natural balance within their ecosystems. Ecosystems were open. Any species could join. They might be recently native to that ecosystem, or they might be aliens brought by happenstance or human hand. The distinction was of no consequence.[9]

This was heresy. Gleason had an uphill battle to get his ideas heard, let alone discussed. 'To ecologists I was anathema', he later complained. 'Nobody would even argue the matter. I was an ecological outlaw.' It took until the 1940s for him to initiate a debate. But by doing so, he had fired the opening shots in a battle for the soul of ecology that has raged ever since. As Michael Barbour of UC Davis summed it up: 'Where Clements saw predictability, uniformity, cooperation, stability and certainty, Gleason saw only individualism, competition, a blur of continuous change and probability.'[10]

By the 1950s and 60s, mainstream ecology was becoming more sophisticated, adopting mathematical theories about how systems work. Eugene Odum of the University of Georgia and others said that ecosystems could be examined according to their nutrient and energy flows.[11] Odum became a key figure in the emerging environmental movement in the 1960s. He argued that ecological thinking about the balance of nature could help resolve a growing crisis caused by 'man's conflict with nature'.[12] Other ecologists followed this path, making big claims for the perfectibility of ecosystems. Of today's generation, Gretchen Daily of Stanford University famously compared ecosystems to aeroplanes. You can remove one rivet from the wing, perhaps a few rivets, without evident harm, she said. But eventually you remove one too many and the wing falls off and the plane crashes.[13]

New ecological science and old ideas about the balance of nature seemed for a while inseparable. But in the 1970s, established systems theory was upset by the arrival of chaos theory. This said that systems, including ecological systems, were not necessarily stable at all. They could be non-linear, subject to abrupt jumps between different stable states, or have no stable states at all. There was no balance of nature. One of the pioneers was Australian zoologist Robert May, who later became chief scientist for the British government and president of the Royal Society in London.[14] He was, like Charles Elton half a century before, a specialist in population dynamics and saw how species went on orgies of boom and bust without any regard for nature's balance. This looked more like the world of Gleason with added mathematics.

The battle between the two camps is partly about the right to claim the legacy of Darwin. Clements and his successors had long argued that natural selection would produce their climax ecosystems. But the individualists and chaos theorists pointed out that, unlike Lamarck, Darwin never believed that evolution delivered any kind of perfect order or natural balance. 'One of the basic tenets of Darwinian evolution', said May, is that 'all animals have the capacity to do more than replace themselves ... The result is that most populations of plants and animals usually fluctuate.'[15] Evolutionary biologist Stephen Jay Gould wrote that 'the Darwinian mechanism includes no concept of general progress or of optimization'.[16] Darwin held that natural selection allowed species to adapt to changing local environments – and that was all. Ecosystems were not perfect or imperfect, good or bad, healthy or sick. There was no goal, no 'climax', no perfect model against which they could be measured. They just were.

Gould, a New Yorker, was one of the great evolutionary biologists of the 20th century. He is most famous for his ideas about 'punctuated equilibrium', holding that evolution wasn't a continuous process, but happened mostly in bursts after major disruptions such as asteroid hits or super-volcanic eruptions. Gould said ecologists were wrong to see anything innately superior about stable ecosystems because disruption was essential to evolution. It provided chances for new species to evolve. Similarly, the notion that 'native must be best, for native has been honed to optimality in the refiner's fire of Darwinian competition' was a 'pervasive misreading of natural selection', and an 'evolutionary fallacy'.[17] Species are not co-evolved to fit niches; they are free to do as they will. Species evolve to get by, not to attain Clementsian perfection. If another species comes along that fits in better, it will supplant the incumbent and become the new 'native'. Or they might carry on together.

Gould pointed out that Darwin had always been fascinated by this chaos of evolution and the role of alien species in it. The great man once wrote to a sailor who had been shipwrecked on Kerguelen Island, a French outpost in the Indian Ocean, to ask if he remembered seeing any seeds growing from driftwood on the beach. Darwin wrote several papers on how seeds crossed the oceans on rafts of vegetation, in the guts of birds or in mud caked on their feet – 'patterns of colonization [that] reflect historical accidents ... and not a set of optimal environments', as Gould put it.[18]

The new generation of ecologists making the case for a rethink on aliens has taken up these themes. They argue that if ecosystems are open and dynamic, then aliens have as much right as natives to join in – and can be just as useful. 'Nativeness is not a sign of evolutionary fitness', says Mark Davis of Macalester College in St Paul, Minnesota, a leading critic of the idea that there is something unfit about aliens. 'Don't Judge Species on Their Origins', he titled one paper. Doing so, he argued, said more about humans and their cultures than about species and their ecosystems. 'Classifying biota according to their adherence to cultural standards of belonging, citizenship, fair play and morality does not advance our understanding of ecology.' Alien species can sometimes be nasty, but so can natives. There is, he believes, a pervasive 'indoctrination' in the science community that builds negative stories about alien species. Aliens are vilified for driving natives out. They are irredeemably 'other'. But this dichotomy has no basis in science, and should be ditched.[19]

If the debate about ecosystems and aliens is, as Davis says, cultural, then it is also political. And the politics is quite toxic. Gould compared Clementsian ecological ideas about perfectible ecosystems with theories that arose around the same time on eugenics, a social philosophy of improving human genetic traits first developed by Darwin's half-cousin, Francis Galton. The parallels are striking. The adherents of both ideas claimed to have their origins in Darwin's theories; both directly translated their scientific theories into a view about how the world should be run; and both had a disturbingly xenophobic world view. Eugenics generated calls for societies to be cleansed of inferior, alien and unfit humans. Clementsian ecology led to a rather similar view of alien species.

The two theories also had common adherents. Many

conservationists of the first half of the 20th century were prominent proponents of eugenics. One was the American Garrett Hardin, inventor of one of the central tenets of environmental thinking, the tragedy of the commons. He advocated forced sterilization of less intelligent people. Another was the Oxford biologist and founder of WWF Julian Huxley, who feared a future world made up of 'the descendants of the least intelligent persons now living'.

*

The big problem for the Clementsian view of the balance of nature is not political, however. It is that it simply doesn't fit reality. Its critics say nature simply isn't like that. Ecosystems are not so much complex and co-evolved machines as the results of accident and chance. Much of what most ecologists have previously regarded as natural, pristine and permanent now seems to be artificial, accidental and recent, with different species constantly moving in and out. Notions of nativeness and alienness have little meaning. Dynamism and change are the norm in nature. Ecosystems that are unchanging may be in trouble rather than in a state of perfection. And that rather messes up our old ideas of humans being in perpetual conflict with the balance of nature. 'We can no longer assume the existence of a static and benign climax community in nature that contrasts with dynamic, but destructive human change', wrote environmental historian William Cronon of the University of Wisconsin, Madison.[20]

To make sense of this, we need a new image of how nature works. Increasingly, ecologists talk not about the brilliance of evolution, or about grand processes of succession leading to climax ecosystems, but about the more prosaic business of 'ecological fitting', a phrase coined by American ecologist Daniel Janzen in 1980.

Janzen is a legendary figure. Born as the Second World War broke out, the son of a director of the US Fish and Wildlife

Service, he is an evolutionary ecologist and conservationist. Since 1963, he has inspired and driven the management of one of the oldest habitat restoration projects in the world: the Santa Rosa National Park in Guanacaste province in northern Costa Rica. There, he pondered a scientific mystery concerning the Guanacaste tree (*Enterolobium cyclocarpum*), an abundant local tree with a huge canopy that grows widely in sunny pastures, providing much-needed shade. It is the national tree of Costa Rica and produces huge numbers of large fruit that contain seeds. But there are no native animals around able to eat the fruit and disperse the seeds. The fruit mostly piles up on the forest floor, unless taken and planted by humans. This is a tragedy for the tree and the park, which is destined to change radically unless the tree's reproductive strategy can be unravelled and revived.

Janzen needed a solution to the 'riddle of the rotting fruit'. Which creatures were the lost dispersers? They would have to be large and herbivorous. His study of natural history discovered that they had been around in the past: ground sloths, bison, camels, and the ancient horses of America among them. But they died out long ago. To revive the forest, he needed to revive dispersal of the seeds of the Guanacaste tree.

His solution was to introduce into the Santa Rosa National Park the modern horse, which had first been brought to the Americas 500 years ago by Spanish invaders. European horses, he found, swallowed the fruit pulp without chewing up the seeds and made an ideal surrogate for the ancient but inconveniently extinct herbivores. For some ecologists, this introduction of an alien into a protected area was ecological heresy. Even if the aim was rehabilitation, it was not 'natural'. For Janzen it was the only long-term way of rebooting the local ecosystem to secure the tree's future.[21]

The discovery opened Janzen's eyes to the notion that alien species can fit in and do a job. That perhaps they have always done this, colonizing new environments, adopting new feeding habits and developing relationships with new neighbours. He called it 'ecological fitting' and wondered if these chance encounters might perhaps be the normal way in which complex ecological communities formed and developed. Perhaps there was more ecological fitting out there than co-evolution.

It would, after all, be easy for ecologists to miss the distinction. They have been trained to assume that cooperation between species, often called 'mutualism', is the result of co-evolution. Janzen wrote in a 1980 paper entitled 'When Is It Co-evolution?' that 'it is commonly assumed that a pair of species whose traits are mutualistically congruent have co-evolved ... For example the fruit traits of a mammal-dispersed seed co-evolved with the mammal's dietary needs. However, it is also quite possible that the mammal entered the plant's habitat with its dietary preference already established and simply began feeding on the fruits.' What looked like co-evolution may very likely be just a happy union. Like horses and the Guanacaste tree in the Santa Rosa park.[22]

Perhaps finding an ecological niche is a bit like falling in love. Romantics may believe they have found the only person in the world who is 'meant' for them, but more likely their partner just happened to be close by and love grew. For pair-bonding, read ecological fitting.

Janzen did not deny that co-evolution happened. He simply said ecologists should not treat it as a default explanation for every relationship between species. Nor should the discovery of such mutuality be used to justify conservation policies that keep aliens out. That would be bad science and could blind conservationists to many opportunities to do their work effectively. For aliens may be essential to reviving ecosystems, as Janzen found in Costa Rica.[23]

If nature is a kaleidoscope of species, constantly reorganizing and adapting, then newcomers will come and go, often in largely random ways. They will fit in as they can, with no more likelihood of doing harm or good than natives. They are not good or bad, nor at a special advantage or disadvantage. They just are. This doesn't mean there isn't any evolution going on. Far from it. There is growing evidence, as we shall see later, that the arrival of new species often creates a burst of evolution and hybridization among both hosts and newcomers as they learn to rub along. But the context is a dynamic, open and unpredictable environment, rather than one in which a fixed group of natives is working to some idealized perfect state. When needs must, species adapt and evolve very quickly to take advantage of their new environments. Many of the most successful have different ecosystem functions in different places – fitting in as necessary.

Stephen Pyne of Arizona State University, a fire specialist, has more opportunity than most to see the dynamic and temporary face of ecosystems. 'Ecologists used to believe nature evolved to create pristine ecosystems of climax vegetation', he says. 'They saw fire as a ruinous interruption to that evolution. But we can now see that fire has a major biological role.' It is essential. 'It shakes and bakes. It frees nutrients and restructures biotas – it takes apart what photosynthesis puts together.'[24]

Nature, like humans, can get stuck in a rut, he says. It needs a clear-out in order to renew itself. A spring clean. Disturbance is essential on all scales, from the local and instant to the global and long-term. Fire is a big part of that, whether in forests or grasslands. It makes space in an individual forest, incinerating dead wood and allowing new saplings to rise up. The American public saw this in 1988 during the massive fire that burned more than a third of Yellowstone National Park. A superabundance of fuel – a result of decades of ruthless fire suppression by well-meaning ecologists intent on preserving their own vision of balanced nature – only required a dry summer and a spark. The result was spectacular, but far from disastrous. The spring clean happened. Two decades later, when I visited the park, the forests were recovered. The main change was to the management of fire. Foresters now allow smaller fires, and sometimes set their own.

Fires, once seen as aberrations, are the new normal. Once seen as destroyers of ecosystems, they are now seen as renewers and enablers. Part of what forests do.

Other leading ecologists have been busy debunking the old orthodoxies. Steve Hubbell of the Smithsonian Institution and UC Los Angeles studied the rainforests of Panama for two decades. He began with presumptions about the forests as climax ecosystems, but eventually concluded that they have no settled arrangement of species. He saw little sign that the current crop of species had co-evolved together. Rather, there was a constant turnover of new species. Even Darwinian forces didn't get much of a look-in. The forest seemed to renew itself almost at random.

The orthodoxy is that when a large tree falls in the forest and a clearing appears, there is a grand battle between species to fill the gap, with the best-adapted winning. Not so, says Hubbell. He never saw it happen. 'The gaps were being occupied largely at random.' Rather than the best-adapted driving out the weaker species, those that moved in were usually just those that happened to be closest at the time.[25] Hubbell has since argued his 'unified neutral theory' at book length. He calls for a rethink not just of how forest ecosystems work, but of how biodiversity happens. The conventional view is that the rich variety of species we see around us is a result of species evolving to occupy myriad different ecological niches. In the hothouse of evolution, specialism requires lots of different species. In fact, he argues, it happens as a result of species being generally so competitively identical, or 'neutral', that many can prosper.[26] Other researchers have argued that the old idea that 'survival of the fittest' means only one species will survive in any niche does not stand up to scrutiny. In fact species often benefit by sharing niches.

The law of Hubbell's jungle is not so much 'survival of the fittest' as 'live and let live'. Most change is random. The result is not optimum – certainly not some pre-ordained perfection – but a workable mish-mash of species, constantly reorganized by a throw of the dice. Some arrangements do not work; but many do. There is room for all in this liberal non-judgmental jungle. Hubbell no doubt exaggerates. His co-author, James Rosindell of Imperial College London, told me: 'The world isn't literally neutral in that way, but we can find out interesting things from its study.' But it does seem that, in real ecosystems, there is rather less ruthless elimination of the weakest than is often assumed, and that ecosystems are generally open – including to aliens.

That may explain new findings about the Amazon rainforest. Hans ter Steege of the Naturalis Biodiversity Centre in the Netherlands reported in 2013 that there are more than 11,000 tree species in the forest. Could they all have a natural niche or vital role in the rainforest system, as conventional ecology suggests? Or did they just hang in there, according to the Hubbell neutral theory? Some tree species did a great deal better than others, of course. More than half of the forest was made up of just 227 'hyper-dominant' tree species. One tree, a type of native palm called *Euterpe precatoria*, had an estimated 5 billion members in the forest. Yet, said ter Steege, such hyper-dominants 'do not have any particular ecological features that stand out'. If Hubbell is right, maybe they just got lucky.[27]

Whatever precisely is going on, this pattern doesn't look like a highly tuned rainforest ecosystem created by co-evolution, or even a very rigorous process of ecological fitting. If the pressure to outperform each other is as strong as commonly assumed, you would expect the most successful species to have obviously superior attributes, and for the rest to fade away. It looks like Hubbell may be right that hanging in there is just fine. No pressure at all.

*

The emerging picture, whether from the neutral theory or ecological fitting or the growing body of evidence of what actually happens inside ecosystems, is that they are much more adaptive and random than is assumed in the old ecology. They are much more open and individualistic and much less collectivist: much more Gleason than Clements. 'Wherever we seek to find constancy, we discover change', says arch-individualist Daniel Botkin. 'Nature undisturbed is not constant in form, structure or proportion, but changes at every scale of time and space.' The idea of nature in balance, in a steady state, or even gravitating towards such a state through a process of succession is false, he says. So too is the idea that change in nature is somehow bad. Environmentalists raise money, legislators pass laws and scientists spend careers trying to freeze nature in a state that is, says James Rosindell, 'neither practical nor desirable'. If nature is always in flux, then trying to stop that flux is anti-nature – and dangerous, because it builds up problems, says Botkin. Just as the 2008 financial crisis resulted from an 'unwarranted faith' in computer models and the reassuringly trouble-free economic forecasts that they produced, so 'the same problem plagues ecology and environmental sciences'.[28] Trying to maintain stability creates instability.

Among many ecologists, there is a move to embrace this new thinking. 'Like most ecologists, I used to assume that alien species were bad', says Jay Stachowicz, a marine biologist at UC Davis. For much of his career, he says, ecological investigations focused on negative interactions between species, like competition and predation. 'Positive interactions have been ignored. I am trying to correct that imbalance, to look at how species co-exist and what happens when invaders show up. We have found that all ecosystems are fairly dynamic and open. By introducing new species we are basically opening up the pool of species for natural selection. And basically species regarded as native and alien are just the same. If you didn't tell me they were outsiders I wouldn't know.'

Many of his fellow ecologists, he says, continue to talk about co-evolution and how native species have grown to naturally 'be' somewhere. 'But these are just-so stories. More and more research shows that species will just fit together. Of course, I can see aliens sometimes do bad things, but natives sometimes do bad things too. It's generally the change we don't like. I don't now believe that alien species are bad per se.'

Amid this radical rethink of old nostrums, some ecologists are willing even to question whether the idea of ecosystems serves any scientific function. An email from one research ecologist brought me up short. 'I don't believe in ecosystems', wrote John Anderson of the College of the Atlantic, a liberal arts college in Maine. He explained: 'People talk about the Gulf of Maine ecosystem as if it were a real thing, a machine with every part inextricably linked to a common purpose. But I ask: where does this ecosystem begin and end? Who are its members?' Arctic terns, he said, spend maybe six or eight weeks in the Gulf of Maine and the rest of the time cruising the South Atlantic. 'Herring gulls go south to Florida. Eels spend most of their lives in the Sargasso Sea before migrating back here. On the other hand, a clam or a mudworm will probably spend its entire life in a single bay within the Gulf. Things that really matter to each species, and each individual within that species, are only tangentially related to anything we would call the Gulf of Maine ecosystem.'

It is, Anderson says, fanciful to think that losing one species will do harm to the system as a whole. 'What matters to a creature is restricted to the interactions it has with a handful of other species.' These interactions 'are very intense and usually very local'. Ecologists may find the idea of ecosystems to be a convenient way of thinking about nature, but 'the word doesn't tell us anything about how nature works', he says. 'A great deal of the environment movement's time during the second half of the 20th century was wasted on a wrong-headed "ecosystem approach" to conservation.'

Ouch. If academic thinking is moving on from an outdated old ecology, where does this leave conservationists? If conservation is predicated on preserving a balanced natural system from being destabilized by man, how should it deal with the discovery that nature is not in balance but constantly changing? That is the subject of the remaining chapters. But, before that, here is a reminder of how nature gets to work, from the purest laboratory imaginable – a new island.

*

Early one November morning in 1963, a crack opened in the bed of the Atlantic Ocean, south of Iceland. Red-hot magma rushed up through the crack, and icy cold ocean water poured down into it. Nearby, fishermen noticed black smoke rising from the sea and a stench of sulphur. For months afterwards, there was a constant submarine battle between the volcano, which spat out more than a cubic kilometre of lava, and the ocean, which did its best to wash it away. Eventually the volcano won, and over three more years of intermittent eruptions, before the eyes of an astonished world, an island formed and solidified above the waves.

The island was named Surtsey, after a mythical Norse fire-eating giant. Its creation was a dramatic affair. Magma had erupted about 120 metres below the ocean surface. Each time the eruptions resumed, the sea boiled and the lava solidified into tiny fragments that formed bright red fountains and sent mushroom clouds of steam and black ash high into the stratosphere. Bolts of lightning shot to the ground from the volcanic clouds. The ash rained onto the newly formed land, often inside giant hailstones. The fallout eventually created a wide black ash plain around a central crater of solid volcanic rock.

It was, some said, like the world being born afresh. Even as early as Surtsey's first spring, in 1964, when Icelandic scientists tentatively set foot on the new island, it was no longer barren. Those first visitors found a fly on the shore, and some seeds, apparently dropped by passing birds. The following spring, they were greeted on the high-tide line by the pretty white flower of a single sea rocket plant, its roots sunk into the ash.

The scientists swiftly declared Surtsey a nature reserve to which they would control access. But from the start, the speed, ingenuity and sheer unpredictability of nature's colonization of Surtsey wrong-footed them.[29] The scientists guessed that the most visible early invaders would be lichens and mosses, brought by sea birds or blown on the winds from neighbouring Icelandic islands. Not so. Instead, flowering plants and grasses took the lead. The sea rocket was no fluke. The species settled in and was swiftly followed by lyme grass and sea sandwort, cotton grass and ferns. It was 1967 before mosses arrived, and lichens limped aboard only in 1970.[30]

A further mistaken assumption was that all the colonizers would be local Icelandic species. Surtsey is only 30 kilometres south of Iceland. If you keep going south, you hit nothing until Antarctica. But it swiftly became a welcome stopping point for birds migrating across the Atlantic between mainland Europe and North America. From the start, visiting geese, ravens and whooper swans brought seeds and insects from distant lands. Snow buntings carried in their gizzards the seeds of bog rosemary from Britain. And while Surtsey's first slugs and earthworms were Icelandic, many of the insects came from mainland Europe.

So far, there are around 60 species of plants on Surtsey, a similar number of lichens, and rather fewer mosses and liverworts. There may be 300 species of insects and hundreds more other invertebrates and microbes. Insects hitched rides on floating tussocks of grass or in birds' feathers. Mites washed up first on a floating gatepost. Spiders came by air, lofted through the atmosphere on silken threads. Some flies are thought to have arrived in scientists' lunch boxes. And a parasitic wasp came as a larva inside the body of a fly that flew from neighbouring islands.[31] But nobody guessed the big winner: the sea sandwort (*Honckenya peploides*). Textbooks say this plant lives on dunes and around the edges of lagoons from Alaska to Ireland to Siberia. It turns out also to like outcrops of magma in the ocean off Iceland. It covers more than 60 per cent of the island.

The first birds nested on Surtsey in 1970. The first chicks hatched just three years after the lava stopped flowing. These early arrivals were all seabirds such as fulmars and black guillemots. They kept to the cliffs, making nests of pebbles, and contributed little ecologically. Not so the gulls. In the summer of 1985, a pair of lesser black-backed gulls built a nest of vegetation and seaweed on the lava flats in the south of the island. They departed, but returned the following year with others and set up a permanent gull colony that now numbers more than 300 pairs.

Some of the scientists visiting the island feared the gulls would destroy the island's nascent vegetation by tearing it up to make their nests. Wrong again. The gulls brought in scraps of vegetation from far and wide. And these airborne flotsam and jetsam, combined with the birds' excreta and the occasional rotting carcass, seeded and fertilized the barren lava and ash. Soon the gull colony created a bright green oasis on the southern tip of the island that has been spreading ever since. Within a decade, there was enough vegetation for geese to come grazing. More recent arrivals include puffins. Willow bushes have become established.

This was an ecosystem created largely by accident. 'One thing led to another and we now have a fully functioning ecosystem on Surtsey', says Borgthor Magnusson of the Icelandic Institute of Natural History, a regular visitor. The plants support insects that attract birds that bring more plants. There was no complex evolutionary adaptation to the surroundings, nor even a replication of ecosystems on neighbouring islands. What came came. Alien or native? Who knows or cares? This was nature doing its thing – ecological fitting in action.

Islands like Surtsey leap up out of the ocean only very occasionally. Most underwater volcanic eruptions either happen at too great a depth or produce too little material to form a land surface that is not instantly washed away. But Surtsey was big. When the eruptions stopped, the island had a surface area of 2.7 square kilometres. And at its summit, the hard lava had solidified into a cliff face 150 metres high.[32]

The nearest modern parallel in terms of geological forces creating a new place for nature to colonize is Krakatoa, an island in Indonesia that was virtually destroyed in a giant volcanic eruption in 1883. The eruption blew the top off the island and destroyed virtually every living thing on it. But within 60 years, Krakatoa had what Charles Elton called a 'rich and maturing jungle'.[33] More than 700 species of insects and 30 species of resident birds had moved in from the neighbouring islands of Java and Sumatra, along with snakes, bats and numerous insects.[34]

Surtsey has excited geographers, who marvel that geological features such as canyons, gullies and land features were created in less than a decade. It has been visited by NASA scientists who say they may one day use it to train Mars explorers in handling the kind of terrain they can expect on the red planet. But Surtsey won't be around for very long. Right now, it is eroding by about a hectare a year. The soft ash is vulnerable to the heavy Atlantic swells. The ecosystem could disappear almost as quickly as it arose. 'We expect that species numbers will continue increasing for the next 30 to 50 years', says Magnusson. 'But after that, they will start to disappear as habitats are lost to erosion.' Seabirds will continue to nest on the cliffs, but most of the unique ecosystem now thriving on the ash plains will be washed away.

Nobody in their right minds will try to preserve this ecosystem. It will depart as it arose, a product of random forces of geology, species migration and ecological fitting. But in that, it is not so different from the rest of the planet.

PART THREE

THE NEW WILD

Alien species may be scary sometimes. But they are nature at its best, and in the 21st century, they may be its opportunity of revival after the damage done to it by humans.

10
Novel Ecosystems

After dusk, the forests of the Caribbean island of Puerto Rico fill with the cry of the native coqui frog. 'Ko-kee', the male frogs croak long into the night. Hence their name. Researchers believe the first half of the call threatens other males, while the second half attracts females. Whatever, after a few drinks, the local islanders reply: '*Soy de aquí como el coquí*' (I'm as Puerto Rican as a coqui). The common coqui (*Eleutherodactylus coqui*) is the unofficial symbol of the island. It turns up in folklore and pop songs, on T-shirts and coffee mugs. A few decades ago, as the island's natural forests were replaced by sugar and coffee plantations, naturalists regarded the endemic inch-long tree frog as being at serious risk. Now the frog is back. But it no longer croaks in the few surviving scraps of native forest, where it has succumbed to a fungal disease. Most of its song comes instead from new woodlands dominated by foreign trees, like the African tulip.

The coqui seems happy there, in what ecologists are starting to term novel ecosystems – composed of new combinations of native species and species introduced by humans, but where the system itself does not depend on humans to keep it going. Is this sacrilege, or is this the future? Is it an ecological abomination doomed to self-destruct, or a model for protecting species and reviving nature in the modern world?

Puerto Rico has a singular history. When Europeans arrived at the end of the 15th century, it was thinly populated by a native seafaring people called the Taino, and still almost entirely forested. Spanish colonists changed that, farming sugar in the lowlands and coffee and tobacco in the mountains. The plantations spread further after the US wrested control of the island during the Spanish-American war of 1898. Sugar production peaked in the 1940s, by which time only 6 per cent of the native forests remained. The island's environment was a mess. With the trees gone, it suffered massive soil erosion and its rivers clogged with sediment. A series of big hurricanes only added to the mayhem. The coqui and many other species went onto endangered lists. The island's growing population was blamed and, fearing a Malthusian apocalypse on the island, US doctors began trialling oral contraception and sterilizing the island's women.

But another future awaited. Export markets for sugar and other commodities dried up. Smallholder farming also faltered as rural people moved to towns to work in American factories. The island experienced economic boom, but agricultural collapse. The forests began to grow back into abandoned fields. Between 1959 and 1974, land devoted to agriculture halved while forest cover rose tenfold to 60 per cent. It was 'proportionately, the largest event of forest recovery anywhere in the world during the second half of the twentieth century', says Thomas Rudel, who studies land use at Rutgers University.[1]

But it wasn't the native Caribbean trees that raced to plant their roots in the former sugar fields. Native species baulked at returning to the infertile, compacted and sun-baked soils on crumbling hillsides that the farmers left behind. Instead, the trees that colonized the former farms were mostly from the island's stock of introduced species. There were more than a hundred of them, the majority imported by Europeans for forestry and agriculture, or as ornamental garden plants. Suddenly the abandoned fields were full of mangoes and grapefruit, avocados, rose apple and, most prominent of all, the African tulip tree (*Spathodea campanulata*). Introduced to the island about a century before as an ornamental tree, it now spread its orange and crimson flowers across the new wild lands.

One man charted the transformation: Ariel Lugo, a local forester and the long-time director of the US Department of Agriculture's International Institute of Tropical Forestry in Puerto Rico. In his 60s now and with a white beard, he has seen the island change hugely. It was, he says, extraordinary, apparently unique and definitely not in the textbooks of either foresters or ecologists.[2] Conservationists were horrified, as their hopes of a natural reforestation were dashed. Instead, they saw an alien takeover. There was, says Lugo, talk of eliminating the aliens and starting afresh. But he stood up for them. His research showed that they were not freezing out the natives, but actually paving the way for their return. The invader trees repaired soils and restored biodiversity. Some provided homes for birds, both native and alien, that subsequently spread the seeds of native plants. With time, many of the more slothful native trees joined the thrusting invaders in the new forests, often now germinated by non-native insects and birds.

The African tulip tree proved a vital colonist, taking over abandoned floodplains in particular. It is now the most common tree on the island. But it is a friendly giant. It is home to the coqui tree frog. It allows light to penetrate to the forest floor. Native species, including reptiles and birds, make up 80 per cent of animal life in the tulip-dominated forests.[3] Seven out of the 60 native bird species were lost during the period of extensive deforestation, but the new forests should prevent further disappearances. Three native finches have been joined by seventeen alien finch species, spreading seeds in the new novel habitats. Parakeets and parrots from other Caribbean islands happily add to the mix. The blue-and-gold macaw is fully established in Puerto Rico, whereas it is on the verge of extinction in its native Paraguay.[4]

Not all natives have revived, it should be said. The two formerly dominant native trees, *Manilkara bidentata* and *Pterocarpus officinalis*, have not so far come back. But without the aliens, says Lugo, forests might not have returned at all.

Instead, two-thirds of the island's forests today contain alien species, but few are dominated by them. The invaders have ended up helping produce what Lugo calls 'beautifully functioning' new forests, with greater biodiversity than the old forests. Eradicating them would be 'fraught with ecological risk'.

Lugo has taken a lot of flak for telling this story of ecological redemption. 'I have been scolded, yelled at, and abused by the conservation priests. Whenever I talk at a conference and give our latest results, I'm met by absolute silence and then, often, hostility from the old guard', he told journalist Gaia Vince.[5] People don't personally hate him, he told me, but 'they are very resistant to accepting the idea of nature's resilience'.

Lugo is unrepentant. A founding member of the US Society for Ecological Restoration, he sees Puerto Rico's new novel forests as a model for the future in many tropical countries. 'Our history of land use means we are ahead of most countries in the cycle of deforestation, degradation and abandonment of land', he says. Old ecosystems will not work in new times, with new climates and new landscapes. Puerto Rico's forests are 'the harbingers of how biota might respond elsewhere to rapidly changing environmental conditions'. He challenges conservationists to encourage their development rather than to try to reinstate what was once there. This, he says, is the new wild.[6]

Conservationists remain confused. Their treatment of the coqui shows this only too well. The frog has the unusual distinction of being both on the IUCN's 'red list' of threatened species and on the same body's list of 100 most dangerous invasive species. It is regarded as at risk because – despite its widespread colonization of the new novel forests across the island – its only 'native' habitat is a small fragment of old Puerto Rican forest, where a fungus is wiping it out. And it is regarded as an official pest, worthy of eradication, in the other place where its population is growing fast – in Hawaii, where it was accidentally introduced in the 1980s.[7]

Ecologists are tying themselves in knots because they refuse to recognize that these novel, hybrid ecosystems are desirable habitats for anything. But as the sun sets across the island, the people of Puerto Rico can simply cherish their new forests. And they can once again carouse with the common coqui.

*

Most conservationists have been reluctant to open their eyes to the discovery of nature's resilience and powers of recovery – still less to recognize the role of specialist colonizers and non-native species in that process. It does not fit their paradigm of how nature is organized. This blinkered approach complicates their wish to protect and revive nature by excluding a wide range of options for rebooting the wild. As Puerto Rico shows, nature often needs aliens and colonists and first-movers. Nature has little regard for conservationists' love of what they see as the pristine. For nature, it matters not a jot where a species comes from, if it does a useful job. If conservationists don't wake up quickly, they risk becoming the enemies of nature rather than its saviours.[8]

This matters hugely because recently messed up and novel ecosystems already dominate huge areas of the planet. There are billions of acres of abandoned landscapes across the world: former forests, degraded pastures, poisoned or radioactive badlands, urban wasteland, and places where farmers simply gave up and walked away. They are where a huge amount of surviving nature lives. They are of necessity the building blocks for nature's revival. Most of these novel ecosystems are former forests and many, unnoticed by conservationists, are returning to bush. Edward Mitchard of the University of Edinburgh found that in recent years, as people move to the cities for work, nature has been reforesting abandoned land in Africa even faster than loggers and farmers cut it down.[9]

The recovery typically starts with specialist colonizers adept at moving in where others fail to tread. Some are old hands, but many are what conservationists see as alien interlopers. Parts of Africa seem to be repeating the experience of Puerto Rico. If conservationists are seriously interested in reforesting Africa, then they should be paying far more attention to helping nature recover in these 'degraded' and abandoned lands. They need to start, says Lugo, by recognizing alien colonizers as part of the solution rather than part of the problem.[10]

We do not have to take it on trust that logged forests and abandoned farmland can recover their former fecundity. As we saw in Chapter 8, almost all forests – even the ones widely regarded as pristine and hotspots of biodiversity – are in a state of recovery from past use by humans. There are no virgin forests out there, says Kathy Willis, director of science at Kew Gardens in London. This discovery, she says, 'has far reaching implications for understanding the resilience and recovery of tropical rainforests' following modern-day human disturbances. It shows that forests can and do recover so completely that ecologists cannot spot the human element. After a few hundred years they are almost indistinguishable.[11] And that in turn shows that what are currently often regarded as worthless 'degraded' forests are actually the vital elements for future revival. As Willis puts it: 'Maintaining such degraded systems in the landscape might be important for building resilience to future disturbances.'[12]

Yet the blinkered thinking persists. Degraded forests, and forests in recovery, are almost everywhere under-researched and under-valued. The most exceptional thing about the new forests of Puerto Rico is that they *have* been researched.

To take one instance of this myopia, many conservation scientists see the logged forests of the giant tropical island of Borneo as virtually a lost cause. Typical was this headline on a press release from the Carnegie Institution for Science in Washington, DC in July 2013: '80 per cent of Malaysian Borneo degraded by logging', it screamed.[13] A new analysis of data from Landsat showed that 'seemingly dense tropical forest cover' was often a mess, 'impacted by previously undocumented high-impact logging or clearing operations'. Collaborator Jane Bryan of the University of Tasmania called it 'a crisis'. The study found that 'very few forest ecosystems remain intact'.[14]

Nobody would suggest that was good news. I have been to Borneo several times and seen for myself the scale of destruction, both from the ground and while flying overhead. The rainforest 'looks like a dog with mange', I wrote once. 'Areas of dense forest are interspersed with large naked patches, ripped out by bulldozers and chainsaws.'[15] But consider this. Most logging is selective, with the valuable trees removed and the rest (or those that don't succumb to collateral damage) left behind. And studies show that forests logged in this way in Borneo retain almost as much biodiversity as many nearby unlogged areas protected in national parks. In among the mess, most species survive, albeit in reduced numbers. David Edwards of James Cook University in Australia found that 75 per cent of bird and dung beetle species remained across a large logging concession covering a million hectares in Sabah, Borneo, even after the entire concession had been logged over twice. Other species would show similar results, he said.[16]

Of course there had been damage and species lost, among which specialist local endemics may be prominent. Old-growth forest will probably contain more species than recent regrowth, or 'secondary' forest. And secondary forests, as in Puerto Rico, are likely to be places where alien species get a grip. But Edwards' point was that 'degraded forests provide important habitat' for the majority of forest species even after logging. Jeffrey Sayer – veteran conservation scientist at the IUCN, WWF and elsewhere – goes even further. He told me that in Kalimantan, the Indonesian part of Borneo, 'biodiversity in logged-over concessions is in better condition than many of the protected areas'. And in the Congo basin of Africa, 'only an expert can really tell the difference between forests in concessions and those in protected areas'. There is, he said, 'abundant evidence' of the conservation value of logged forests.

Daisy Dent, a tropical forest researcher at the University of Stirling, says that globally 'there is now a larger area of secondary and degraded tropical forest than there is old-growth forest'.[17] This, surely, is too much biodiversity to be ignored. 'Secondary forests of all ages should be protected. They are a hugely valuable safety net for biodiversity', says Robin Chazdon of the University of Connecticut. 'And like a good Bordeaux, the value of secondary forests is expected to increase over time.'[18] Edwards' colleague William Laurance agrees. He told me: 'Logged forests in the tropics are too vast, vulnerable and important to ignore, given their large conservation potential. It is vital that we recognize their key role for conserving tropical nature.'

Yet these places are rarely recognized as being of conservation value. No green group stands up to protect secondary forests, even though they are the most abundant habitat for wildlife in the modern tropics. They receive little more attention than the forests that have been clear-felled and converted into plantation monocultures of acacia or eucalyptus. They are routinely left off maps of the world's surviving forests, dismissed as 'degraded' and valueless, says Douglas Sheil of the Norwegian University of Life Sciences. Worst of all, conservationists often actively encourage their conversion to agricultural plantations, in a misguided attempt to protect other forests from conversion.

The Washington DC-based World Resources Institute is among those groups supporting a billion-dollar Indonesian government plan to save 'pristine' old-growth rainforests by encouraging palm-oil producers to take over 'degraded' forest land. But Laurance says that a lot of the 36 million hectares of Indonesia that has been designated as 'degraded' – an area larger than Germany – could be almost as rich in species as the old-growth forests. 'Preventing degraded forests from being converted to oil palm should be a priority of policy-makers and conservationists', says Edwards. But it isn't. By elevating the conservation status of supposedly pristine parts of nature, and disregarding the rest – the new wild – conservationists end up complicit in forest destruction and biodiversity loss.

*

Henry Gleason's ideas about species being individualistic lead logically to the conclusion that novel ecosystems are going to become an increasingly important part of nature. If ecosystems are not 'super-organisms' or grand associations of species drawn inevitably to a particular 'climax', then they can assemble anywhere, however species like, and in whatever form works. They have no pre-ordained template. No guest list. The spread of novel ecosystems, apparently functioning perfectly successfully, is of itself strong evidence that Gleason was right. As enthusiast Richard Hobbs of the University of Western Australia puts it, novel ecosystems are 'the new ecological world order' and 'in some cases provide the only habitat available for species of conservation concern'.[19]

Even Charles Elton, the founder half a century ago of the science of alien species, recognized that the pristine world, the old wild, was gone. He was mournful rather than fearful. 'The wilderness is in retreat', he wrote at the end of his book. 'The balance of nature does not exist and perhaps never has existed ... We must grow accustomed to the idea that its plants and animal populations will have changed their composition and their intricate structure and relationships.' In California, he noted, Asian red scale insects lived widely in groves of citrus trees from South-east Asia, and did specially well where Argentine ants ate their natural enemies. All were aliens, but had made themselves at home conducting a productive relationship in a foreign land. We should, he said, embrace that diversity. Conservation 'should mean the keeping or putting in the landscape [of] the greatest possible ecological variety. I see no reason why the reconstitution of communities ... should not include a careful selection of exotic forms.'[20] That is not a bad manifesto for novel ecosystems – from the man who triggered our fear of aliens.

What was once thought of as pristine is mostly far from virgin. But equally, much that has been written off as messed up nature still has great conservation value. Conventional conservation, said the late Stephen M. Meyer, MIT political scientist, operates 'on the grossly mistaken belief that we can halt ongoing extinctions, [which] fuels our preoccupation with saving relics and ghosts'. What conservationists should really be doing, he said, is 'turning our attention to the new assemblages of organisms that are emerging' as a result of our activities.[21]

Slowly this is happening. Puerto Rico is far from unique. Many other islands – once thought of as casebook studies in the dangers posed by aliens – are now becoming the crucibles for understanding the value of alien species in novel ecosystems. Mauritius, a volcanic island in the Indian Ocean settled by Dutch sailors in the 17th century, still contains almost 90 per cent of its 765 recorded native plant species, while another 730 introduced species have become naturalized. We may mourn the loss from the island of its giant fruit-eating animals, the flightless dodo, tortoises and giant lizards. All were driven to extinction by the Dutch sailors. But at least now there are alien substitutes to do the housekeeping. Wildlife managers have introduced giant tortoises (*Geochelone gigantea*) from Aldabra 2,000 kilometres away to spread the seeds of native trees. The newcomers have, says science author Sharon Levy, 'rescued at least one endangered plant – *Diospyros egrettarum*, a species of ebony endemic to Mauritius'.[22]

Other islands show similar commendable novelty. Take the Seychelles, another vacation idyll in the Indian Ocean. The islands were largely deforested by French and British colonists who replaced the trees with plantations of spices like cloves, nutmeg and cinnamon (*Cinnamomum verum*) from India. But the farmers stopped bothering with their fields when an international airport opened and tourism took off in the 1970s. As on Puerto Rico, the forests grew back. More than 90 per cent of the island's 325 known native plant species were still around, and they began to move back onto their old terrain. But what is emerging across the islands is a novel forest ecosystem, in which ancient endemics grow side by side with the spice trees. Cinnamon proved the best colonist and is now the most abundant tree on the lowlands.

Some conservationists want to get rid of the cinnamon and other alien trees, to give the natives a chance to flourish once more. But James Mougal of the Seychelles National Park Authority says cinnamon and the others represent nature's best shot at a resilient new ecosystem. They provide excellent forest cover and leaf litter for native amphibians, snakes, insects, snails and plants. The endemic Seychelles fruit bat roosts primarily in alien trees, whose fruit and seeds are a major food source. In any case, eradication would likely fail. Mougal's compromise is to re-establish more patches of natives among the cinnamon trees to ensure they have the best chance of taking their place in the new novel ecosystem. That seems smart.[23]

*

The recognition that novel ecosystems are the future – the new wild – is often painful for conservationists. And nowhere is that more true than on the volcanic archipelago of the Galapagos Islands, 1,500 kilometres west of Ecuador. The islands are famous for conservation. They are the home of giant iguanas, sea lions, numerous endemic bird species including the only tropical penguin, and, until his death in 2012, Lonesome George, the last surviving individual from the Pinta Island sub-species of giant tortoise. The islands are also important for science. It was after visiting here that Charles Darwin developed many of his ideas on evolution. Probably more time and money has been spent on conserving the biodiversity of these remote islands than anywhere of comparable size on the planet.

So it came as a shock when in 2011 Mark Gardener, after two decades working to preserve the islands' unique species, seemed to haul up the white flag. The head of restoration at the Charles Darwin Research Station, which is in charge of most conservation activity on the archipelago, told *Science* magazine: 'It's time to embrace the aliens. As scientists and conservationists, we need to recognize that we've failed. Galapagos will never be pristine.'[24]

The Galapagos Islands are volcanic lava flows that surfaced in the past 5 million years. The climate is cool and dry, since for most of the year the islands are bathed in the cool Humboldt Current coming up from Antarctica. There are frigate birds, boobies, petrels, mockingbirds, flightless cormorants and other endemic birds. There are a wide variety of iguanas, and the finches – with different beaks to eat different foods on different islands – that so fascinated Darwin.

The islands were probably uninhabited before Europeans arrived. Aliens came with them. The first black rats hopped off warships in the 17th century, and began eating the eggs of both endemic birds and the giant tortoises. Goats, cats and others soon mutinied from whaling ships that were visiting to stock up with tortoise meat. The population of alien species exploded more recently as local fishers and others began to drop anchor regularly, followed by a permanent population to service the 200,000 tourists that visit every year. Now there are more than 500 introduced plant species – almost as many as there are native plants – along with 500 invertebrates and 36 vertebrate animals. Blackberry bushes, originally from Asia, cover 30,000 hectares, taking over from the forests of *Scalesia* trees in the uplands of Santa Cruz and other islands.

In 2000, a ten-year project, funded with $19 million from the World Bank's Global Environment Facility, set a goal of the 'total control of invasive species' on Galapagos. Not eradication, but control. Altogether there were 43 projects targeting 35 species of invasive plants, animals and invertebrates. But a decade later, just nine of the projects had achieved their targets.[25] Restoration of native plant cover had been achieved on less than 200 hectares, said Gardener. And even those successes were 'not stable' and would 'require continued high-level intervention' to keep the invaders from returning. Efforts to uproot the alien guava and blackberry bushes had been worse than useless, he said. The digging created more of the disturbed ground that invasive species so enjoy. Eradication of aliens, Gardener concluded, 'is not a viable tool'. The white flag was raised.[26]

But all is not lost. 'The Galapagos campaign that Gardener deems a failure is rife with success', insists Daniel Simberloff.[27] Around a quarter of a million goats have been eradicated from both Santiago and Isabela Islands in campaigns involving hunting with dogs, shooting from helicopters and the mass release of hundreds of sterilized females to distract the males.[28] The recovery of vegetation following goat eradication has brought the endangered Galapagos rail back from the brink of extinction on Pinta Island.

Simberloff has a point. Despite the invasions, the Galapagos Islands have lost very few species. The sub-species represented by Lonesome George is a high-profile exception. What Gardener wanted to do was call off the 'war' on aliens. Instead, he wanted to sue for a permanent peace, coming to an accommodation with the aliens by working out how they might fit in while still protecting what is most worthwhile about the old guard. In other words, he wanted to start work on engineering 'novel ecosystems' on islands more synonymous with conservation than any other on the planet. Simberloff argued that this was good management, even if short of eradication. The difference was more rhetorical than real. Finally, in the Galapagos, there seemed to be a meeting of minds. If this is the way forward for Galapagos, then surely it is the future of conservation across the planet.

*

This is not a counsel of despair, even for those who want a revival of nature as near to what we had in the past as possible. The story of the 'pristine forests' is encouraging. Forests can regenerate in a few centuries so that even experts cannot tell if they are pristine old growth or secondary regrowth. And there is growing evidence that most other ecosystems can do the same. The idea that modern man is destroying nature for ever is simply unsupported by the evidence. Even a few decades will see much of it righted.

Holly Jones and Oswald Schmitz of Yale looked at 240 scientific studies of what happens to damaged ecosystems if humans walk away. They found 'startling evidence that most ecosystems globally can, given human will, recover from very major perturbations'. More than a third showed full recovery; as many again showed partial recovery; and only 67 showed no recovery. And it did not take the 'centuries and millennia speculated previously'; it usually took about 40–50 years for forests, where trees have to regrow, and less for other ecosystems.[29]

That time frame may not bring back the full range of species. Often, newcomers will move in. But the authors said their evidence 'does not support gloomy predictions', among which they included the bleak assessments of the prestigious Millennium Ecosystem Assessment. The MEA, headed by Bob Watson, chief scientist at the World Bank and others, declared in 2005 – without, Jones and Schmitz noted, providing a reference – that 'once an ecosystem has undergone a nonlinear change, recovery to the original state may take decades or centuries and may sometimes be impossible'.[30] Of course, the MEA was right about the 'original state'. Nothing ever goes back to its original state. But rich ecosystems would usually reappear. There was, Jones and Schmitz concluded, 'much hope that humankind can transition to more sustainable use of ecosystems'.

This story of ecological recovery is hugely good news from a major analysis of hundreds of studies. And yet when I checked Google Scholar four years after publication, Jones and Schmitz had received a paltry 59 citations. Do scientists simply not want to know the good news? Jones thought not. 'In general, there is a tendency to be pessimistic about ecosystems' likelihood of recovery, even when most of the data show it is possible', she said. 'When we find recovery, we may first think, "Did I measure the right thing? Am I missing something?" rather than saying, "Right, this is an example of a success".'

Some fear that accepting the value of novel ecosystems and the alien species within them suggests an 'anything goes' approach – 'tantamount to giving up the good fight for conservation', as Andrew Light of the George Mason University puts it.[31] Logging? Why not? Toxic waste? Some species will like it. Farming? No worries: when the ploughs are put away, nature will always come back. But it is not a call to let rip. It simply offers hope and realism. It is surely folly to ignore the conservation potential of degraded and logged-over forests, or of abandoned farmland, simply because they don't match our idealized expectations of the pristine. Conservation in the 21st century requires an open assessment of what might work. Not a sullen retreat into blinkered orthodoxy.

11

Rebooting Conservation in the Urban Badlands

Peter Shaw clambered over a huge pile of ash outside one of Britain's larger power stations. He trod carefully to avoid the profusion of orchids, tested the thin smear of soil for signs of the first earthworms, and lifted an abandoned sheet of metal to uncover a grass snake and a slow worm. Shaw was exploring one of hundreds of forgotten biological treasures in Britain's industrial badlands. They are, he says, the extreme case of our new novel ecosystems, and ecological resources of increasing importance in our crowded landscapes. 'Planners dismiss waste tips, old industrial sites and similar places as ripe for redevelopment, but they often support more scarce wild species than farmed land', says Shaw. 'They have one in six of the UK's rare insects, for instance.' Nature persists, even flourishes, in the most unlikely, most damaged and apparently least natural environments. We should value them as much as any rainforest, says Shaw, a soil biologist at the University of Roehampton in London.

The nation that gave birth to the industrial revolution is now the home of a remarkable wildlife revival on derelict land. Nature is taking over ash heaps, old chemical factories, former oil refineries, railway sidings and metal mines. Rare native species find in such places bizarre man-made habitats that do not exist elsewhere. As do exotic species, some of which move in specially. Yet few conservationists have caught up. They rarely protest when these industrial hotspots of biodiversity are built over. Instead, they call for such places to be developed in order to keep the bulldozers out of green fields – green fields that are often biologically sterile. Most field guides to British wildlife simply never mention these badlands. This casual indifference to nature away from where we expect to find it is perverse.

Shaw and I were standing on a heap of pulverized fuel ash at Tilbury, on the estuary of the River Thames east of London. Shaw had been checking nature's progress on a series of observation plots he has returned to regularly for more than twenty years. This time he found that the small yellow flowers of bird's-foot trefoil were doing well. The plant likes the alkalinity, the infertility and the high levels of the metal molybdenum in the ash. They help it fix nitrogen from the air and eventually to create soil. Shaw also logged sea buckthorn, yellow-wort, melilot and mouse-ear hawkweed, before checking out what looked like cannabis but turned out to be American willowherb. The place is a treasure trove. Orchid lovers have found dozens of hybrids here. Entomologists counted 656 different invertebrate species, including a dung beetle thought extinct since the 1920s. Bees are busy. 'But no earthworms yet; it's still too alkaline', said Shaw.

There are bigger fauna, too. At night, barn owls swoop over the ash heap from their new home in an abandoned part of the power station. Deer somehow make their way here through the maze of roads leading to a nearby container port. We stumbled over rabbit holes. 'Rabbits love the ash', said Shaw. 'It is soft on their paws, but holds its shape when they burrow into it.'

Coal-fired power plants produce millions of tonnes of ash a year. Half goes for construction, and the remainder is dumped in old lagoons or giant waste mounds. When not encouraging building on the mounds, planners like to cover them with soil and plant trees in the name of 'ecological restoration'. But Shaw believes the piles should be left as they are. When that happens, they become brief but brilliant oases of biodiversity, as at Tilbury. After a few decades, more regular chemistry takes hold in the emerging soils and the punkish and lawless landscape gives way to middle-aged woodland. But while it lasts, there is nothing like it. Such places often have no equivalent in more natural settings.

The shoreline of the Thames estuary is dotted with old industrial sites that combine plenty of space and not too much interference. They provide rich pickings for 21st-century wasteland safaris. A couple of kilometres from Tilbury, Shaw and I headed past the Lakeside shopping mall, skirted a Procter & Gamble detergent factory, and pushed on beneath a humming electricity pylon. We spotted the pretty flowers of stonecrop, dog rose and buddleia, which was first introduced to England from the Caribbean 300 years ago and named after a plant-collecting rector from Essex, Adam Buddle.

Our destination was West Thurrock Lagoon, another leftover from a demolished riverside power station. Its saltmarsh, bare ground and scrubby meadows have recorded more rare species than almost any other site in Britain, including 36 different types of bees and many rare invertebrates. It is one of only two known UK sites for the distinguished jumping spider (*Sitticus saltator*). Or it was. As we clambered over the graffiti-covered sea wall to get our first look, Shaw gasped. Much of the lagoon had dried out since his last visit. The place was overgrown with hemlock. This was, he said, almost certainly bad news for the distinguished jumping spider, whose only other British home is another alkaline former industrial site on the other side of the estuary that has been earmarked for a theme park.[1]

Further downstream at Canvey Island, a derelict oil terminal by Holehaven Creek has more species per hectare than any British nature reserve. Amid old bikes and broken concrete, burned-out cars and wind-blown litter, the 1,300 species include the shrill carder bee and four more of the country's rarest bumblebees; 300 species of moths; a weevil (*Sitona cinerascens*) not seen in Britain for 77 years until it was rediscovered here in 2005; and the scarce emerald damselfly, also once thought extinct in Britain. It is also the last outpost of the Canvey Island ground beetle.[2]

'Brownfield sites are as important for biodiversity as ancient woodlands, yet we are encouraging people to build on them', says Matt Shardlow of the UK conservation organization Buglife. 'It's the combination of habitats that is so rare. There are very bare areas, basking places, short grasses, sallow scrub, sand dunes, poor land, rich land, and bits of wetland.' Trail-biking youths and illicit bonfires ensure that trees never take over.[3] Feral urban Britain turns out to be a wildlife paradise.

*

Many old industrial processes have left behind unique habitats. In what used to be called the Black Country of the English Midlands, limestone slag heaps left by early blast furnaces now harbour rare meadow grasses. In 19th-century Britain, the Leblanc process for manufacturing soda from common salt produced huge steaming piles of blue alkali paste that contained quicklime, calcium sulphate and unburned coal. It scorched the earth and smelled of bad eggs. Factories in north-west England dominated world soda production and at one stage generated annually hundreds of thousands of tonnes of this toxic brew. The piles have now weathered to limey soils where vast swathes of orchids grow at places with splendid local names like Mucky Mountain, Lower Hinds, and Shaw's favourite, Nob End in Bolton, which covers 5 hectares where the rivers Irwell and Croal mingle. Nob End harbours rare plants like carline thistle and blue fleabane, and has eight types of flowering orchids, including several hybrids that do not fit textbook descriptions.[4]

I was pleased to discover that another of Shaw's haunts is the Brockham lime works in the Surrey hills south of London, a favourite picnic haunt of my family for decades. I knew about its orchids and the rare silver-spotted skipper butterfly but not the 24 species of *Collembola* springtails, including one, less than a millimetre long, that Shaw discovered there back in 1999. It is found at just one other place in Britain. It lives on. But no such luck at Gargoyle Wharf, an oil depot a kilometre or so from where I live in south London. After the depot shut, botanists found more than 300 species of flowering plants there. It was, Nick Bertrand from the London Wildlife Trust told me, 'one of the most fantastic sites I've ever been to'. But big-league conservation groups said it was a brownfield site and left it to developers, who built apartment blocks and replaced the unique flora with 'award-winning landscaped gardens'.

This environmental philistinism has come about because all of us – townies and country-dwellers, planners and environmental campaigners, developers and nimbys – have come to believe that cities are environmental deserts, while the countryside is the only proper place for wildlife. The truth is often the opposite. Many green fields that the countryside lobby wants to protect are ploughed-up, pesticide-soaked ecological deserts. Many urban sites they want built on are unique places, extreme novel ecosystems that are refuges for nature's rare, bizarre and itinerant.

Post-industrial landscapes are helping turn our picture of nature on its head – or they would if we were taking more notice of what is going on in their nooks and crannies. With most of the British lowland countryside now prairie farmland and monoculture pasture, wildlife is being squeezed out of where we expect to find it. By contrast, when industry abandons land, the open spaces and weird niches it leaves behind are full of unusual chemical and physical habitats, and ripe for colonization. Everything is up for grabs, and wildlife grabs it. England's brownfield sites are 'the new lowland heaths', says bugs expert Peter Harvey of the Essex Field Club. They are messed-up places that have 'much more in common with the historic wildlife-rich countryside than the intensively farmed modern version'. They 'routinely support more scarce wild species than farmed land'.[5]

That is why Europe's largest population of great crested newts (*Triturus cristatus*) – about 30,000 of them – is happily ensconced in ponds that now occupy old brick pits outside

Peterborough in eastern England. And why, at least until redevelopment plans go ahead, Britain's top site for nightingales (*Luscinia megarhynchos*) is a military junkyard, littered with abandoned munitions stores, sentry boxes and an anti-aircraft gun emplacement, on Lodge Hill near Gillingham in north Kent. Amid the junk, where trainees from the Royal School of Military Engineers once practised driving bulldozers, are patches of scrubby woodland where 90 pairs of one of Britain's favourite birds sing their hearts out.[6]

This Cinderella ecology isn't so new in Britain. The last windfall of sites for rare natives and exotic invaders happened after bombs dropped on London and elsewhere in the Second World War. The profusion of unexpected species that populated the craters was so great that it was rumoured they had been dropped with the bombs as biological weapons of war.[7] The Moroccan poppy and the American willowherb were both first spotted in Britain in the remains of bombed-out buildings in the City of London, and subsequently spread across Britain. Those were good times for thorn apple from North America, and rosebay willowherb from the Yukon, which was nicknamed 'bombweed' by Londoners. Some were newcomers, but many were old arrivals. The daisy-like gallant-soldier (a corruption of its Latin name *Galinsoga parviflora*) came to Kew Gardens from Peru in the 1790s but proliferated unexpectedly in the bomb craters.

All this was documented by City of London banker and amateur botanist Ted Lousley, who reported on more than 250 plant discoveries in his book, *The Natural History of the City*, published in 1953.[8] He treasured both alien finds and the hybrids produced by illicit liaisons between aliens and natives. He often took samples and planted them in his own garden in south London, which became for a time the country's largest private herbarium.[9]

Besides the craters, his favourite places to search were railway sidings, rubbish tips, graveyards and market gardens, where horticulturalists used the fine-shredded waste from the wool industry, known as shoddy, as a cheap fertilizer. There were rich pickings here because the shoddy carried with it many plant seeds from foreign climes that had become entangled in the fleeces. Over the years, he found more than 500 species trapped in shoddy, including Mediterranean plants that Merino sheep from Spain had taken to Australia and New Zealand in the 19th century and that later came to Britain in their fleece.[10]

Lousley's jottings started the botanical study of brownfield Britain. His successor was northern lichen-hunter and botanist Oliver Gilbert of Sheffield University.[11] In 1993, Gilbert wrote the first government report anywhere in the world on the ecology of derelict urban sites, which he reckoned contained more of the nation's rare insects than either ancient woodlands or chalk downs. 'At least 40 invertebrate species are wholly confined to brownfields', he wrote. Rare wasp and bee species love abandoned quarries. The glow-worm mostly lives now in old walls. 'Butterfly species may now be more likely to be found in suburban areas than in rural areas', due to the sheer variety of habitats in suburban gardens and parks. Less surprising, perhaps, is that bats inhabit buildings and tunnels. Indeed, many now roost almost nowhere else.[12]

Nature's go-getters, carpetbaggers and vagabonds end up in these badlands. Michael Crawley of Imperial College London ranked various types of British habitats according to how 'invaded' they were by alien flora. The top habitat for aliens was waste ground, with 78 per cent of the flora made up of aliens, followed by conifer plantations (56 per cent) and walls (46 per cent).[13] The warm and dry microclimates in open spaces typical of brownfield sites are often the first stopping places for migrating Mediterranean species. But in these novel ecosystems, the distinction between natives and aliens is particularly meaningless. No species is 'native' to a chemical works or ash heap.

We should be wary of pushing the full panoply of environmental protection onto these places, however. Not least because it wouldn't work. Brownfield sites do not have to be protected for the long term. Their charms and biodiversity credentials are almost always ephemeral. No ecosystem is permanent, but most brownfield sites, as homes of new colonists on often bare ground, are changing faster than most. They are nature at its most dynamic, but also most temporary. Preserving them for posterity is a contradiction in terms. But, while they are in bloom, we should treasure them.

*

Around the world, nature is moving to the cities. 'Ecological novelty pervades the urban environment', says Michael Perring of the University of Western Australia.[14] Gardens and cemeteries, abandoned industrial areas, transport corridors and even suburban trash cans are all grist to nature's mill. Sometimes cities provide specialist habitat. Buildings and bridges in cities from Budapest and Florence to Brussels and New York provide substitute cliff roosting sites for birds of prey such as peregrine falcons (*Falco peregrinus*). For a decade now, I have enjoyed watching the fastest birds of prey in Europe swooping on city pigeons from their nests in the turrets at Chichester cathedral on the south coast of England. They seem to like it as well as their 'proper' sea-cliff habitat.[15]

More often, cities are irresistible food sources. Australia's once-rare grey-headed flying foxes found so much food in Melbourne that a colony of 30,000 of the bats has formed there in the past two decades. Raccoons (*Procyon lotor*) are joining others who forsake the rural life for city scavenging. They are smart enough to negotiate any American urban obstacle course in search of a meal. They are 'able to squeeze into locked garages, open secured garbage cans, unzip tents, and pry up lids on Tupperware', wrote one blogger after watching a PBS documentary.[16] Mile for mile, there are five times as many raccoons in American suburbs as in the surrounding countryside.

Oddly, many species find cities safer than the countryside. The coyote (*Canis latrans*) lived mainly in the south-western states of the US until the 20th century, but then headed for the cities. In the absence of hunting, their survival and reproduction rates are higher there. There may now be 2,000 coyotes living in the suburbs of Chicago, navigating the city's highways by night with rarely a mishap. Los Angeles, New York and Boston also have substantial populations.[17] Ecologists say they are the new top predators on the mean streets. Like foxes in the UK, they are in part fleeing human foes in the countryside. But while British foxes no longer fear showing themselves, coyotes keep to dark places and go out mostly at night, 'quietly conquering urban America', as the *Economist* put it.[18] Golden-headed lion tamarins, squirrel-sized monkeys, came out of the disappearing coastal forests of Brazil and found a new home in the suburbs of Rio de Janeiro. That, as James Barilla of the University of South Carolina points out, makes them both endangered and invasive.[19]

Many species that traditionally get on well with humans have become convinced urbanites. The house sparrow (*Passer domesticus*) seems to have been with us at least since we started farming. Rarely found anywhere remotely wild, it sticks with humans, their landscapes and buildings. The relationship has served the bird well. It is probably the most common bird in the world. Only the chicken comes close. A flock lived for several years inside Heathrow Airport's old Terminal 2, feeding off crumbs from the snack bars. Sparrows will even join us underground. Yorkshire coal miners at Frickley Colliery found a nest 600 metres down at the bottom of a shaft in the mid-1970s. The birds stayed for three years.[20]

In recent years, sparrows have been in decline in some of their favourite urban environments, with numbers halving in Britain since the 1970s. Nobody is sure why. Theories range from unleaded petrol and mobile phones to our urban tidiness. But the fact that, after thousands of years, they seem to be finding us uncongenial is worrying.[21]

There are few such fears for the wild boar (*Sus scrofa*). It is another old friend that has certainly not lost its love of human habitat. We domesticated it 9,000 years ago. Since then it has ranged the Old World, from Japan to Britain and Indonesia to the Atlas Mountains of Morocco. It has often been our calling card. The Polynesians took it to Hawaii, the Spanish to Florida, and the English to New England and Australia, where some 20 million now roam free. The US has some 6 million of them. The wild boar is not a fussy companion. It will roll around in mud or dig up golf courses. It will sweat it out in Texas or the Borneo rainforest, but produce its own steam in the forests of Siberia. It will eat kitchen scraps, mushrooms, snails, turtle eggs, live birds or rotting carcases.

Such *Homo sapiens*-loving species adapt to city dwelling in interesting ways. Grey squirrels get more aggressive and daring. Many birds sing louder and move up the scale, singing higher notes that are less likely to be drowned out by the rumble of city traffic.[22] Pigeons make what appear to be regular planned journeys on the London Underground, saving their wings and energy as they commute to get to food supplies or return to their nests.[23]

Studying all this has long been a backwater of science. But a new generation of researchers is catching up, with resources like the 'Nature of Cities' blog, founded by New York environmental scientist David Maddox.[24] Britain's University of Bristol has set up an urban pollinators project, tracking down urban wildflower meadows.[25] But many governments are hostile. They see cities as hotbeds of alien invasions, disreputable ecosystems and species that just shouldn't be there. The European Union, which has passed new legislation to force governments to act against aliens, reported in 2013 that invasive species 'threaten urban environments'. It never quite explained why.[26]

The IUCN published its own study of 'invasive alien species' in urban Europe, highlighting the presence of American bullfrogs and Canada geese in Flanders, rabbits in Helsinki, hogweed in Estonia, New Zealand flatworms in Scotland, the raccoon in Berlin, the Hottentot fig in Dublin and Indian parakeets in London. The charges included spreading disease, damaging monuments and triggering allergic reactions.[27] They also 'take over resources and space from the indigenous species', said Chantal van Ham, the IUCN's European programme officer.[28] But do they? Or do they expand biodiversity and create opportunities for others?

Urban parks, gardens and allotments are increasingly valuable spaces for wildlife of all sorts. In Britain, domestic back gardens cover up to a quarter of most urban areas. A study found that 70 per cent of their flora is foreign.[29] Does that make them a 'threat to urban environments'? Often their lively mix of natives and aliens *are* the urban environment. Allotments are the best habitat in Britain for bees. According to the government agency Natural England, urban gardens, even if dominated by aliens, are vital habitats for many native species, from the common frog and the song thrush to the hedgehog.[30]

Transport links can be equally important novel ecosystems for wildlife. Railway routes have been valuable ways for alien species and others to move around. Highways, too. Conservationists usually see roads as barriers to migrating wildlife, because they fragment the landscape into small pieces and can trap the unwary. But highway verges can also be migration corridors. In the 1980s, the British government successfully recruited the M40 roadside verge between London and Oxford to link two protected woodlands. Invertebrates such as butterflies could not make the journey using the surrounding countryside because the fields were full of agricultural chemicals. So wildflowers and blackthorn bushes were seeded beside the roaring traffic. The plan worked, and 25 butterfly species, including the rare black hairstreak, duly colonized.[31]

*

Europe's most remarkable brownfield site – an area the size of Luxembourg – is the exclusion zone around the stricken Chernobyl nuclear reactor in Ukraine. With humans banished since the catastrophic nuclear accident there in 1986, creatures of all sorts have come flooding back to this once-busy landscape of farmland, villages, urban areas and forests. This came as a huge surprise to almost everyone. In the immediate aftermath of the world's worst nuclear power accident, a cloud containing radioactivity equivalent to twenty Hiroshima bombs blew north from the burning power station, heading over the border from Ukraine into Belarus. Some areas received massive radiation doses. An area known as the 'red forest' turned rusty brown and died. After the accident, Soviet authorities removed all humans living within 30 kilometres of the reactor. The risk of the lingering radiation causing cancers and other diseases was very real. But the immediate damage to the nature they left behind was mostly much less drastic. While the area still sets off Geiger counters, nature has made a huge comeback. Native species have revelled in the absence of humans, and many new species have moved in, too.

Today, radioactive wolves (*Canis lupus lupus*) patrol the streets of Pripyat, once a town of 50,000 people and now the largest ghost town in the world. Strontium-stuffed mushrooms flourish in the surrounding marshes. Mice scamper around the abandoned power-station reactors, and wild boar root in the caesium-soaked soils. British radiation physicist and keen angler Dave Timms caught a 30-kilogram catfish in contaminated Chernobyl cooling ponds, where altogether 38 fish species live.

For some wildlife, the radiation may make their lives shorter. But nature overall is doing fine, living off the fat of the land. Top carnivores like lynx, eagles and wolves do especially well. There are also moose, beavers, badgers and otters. Migrating birds drop by as if nothing had changed. Black storks, green cranes and white-tailed eagles are breeding. Deer shelter from winter storms in derelict country cottages. Rich grasslands and pine forests are moving in on the old collective farms. Sergey Gaschak, a Ukrainian naturalist who works for the International Radioecological Laboratory in Slavutich, the city built to house refugees from the exclusion zone, told me: 'There are more opportunities for wildlife. The villages and towns have more diverse conditions than surrounding landscapes, with buildings, ponds, gardens and different kinds of vegetation.' Only pigeons and rats, which once relied on human leftovers to flourish, have failed to prosper.

Visiting Western scientists agree. 'Most people think of the zone as a post-Apocalyptic wilderness, either occupied by two-headed monsters that glow in the dark, or completely empty', says Jim Smith of Portsmouth University. 'But from the wildlife point of view, the disaster has been beneficial, because it forced people out. Wild animals rarely die of the diseases of old age. Wildlife in the Chernobyl zone is now more abundant and diverse than before the accident.'[32] Cham Dallas of the University of Georgia found that Chernobyl mice are more radioactive than any creature ever found before in the wild. Yet they seem virtually untouched by the experience. Another team from Texas Tech University, led by Ron Chesser, found that genetic variation among voles in the exclusion zone reflected nothing more than natural variability. This 'failed to support' arguments that there were mutations as a result of the accident.[33]

In Belarus, where the radioactive fallout was greatest, scientists swiftly saw the conservation potential of an area from which people were excluded. Two years after the disaster, they set up the Polesye State Radiation Ecological Reserve to protect wildlife in the zone. It is, by some measure, the largest area in Europe set aside exclusively for nature. Fearing radioactive water flowing out of the zone into rivers like the Pripyat and Braginka, the Belarus authorities dammed up drainage channels. The result is the most radioactive wetland in the world, containing a gloriously diverse assembly of black grouse, partridge, roe deer and much else. Ukraine announced in 2014 that it plans to create its own biosphere reserve in its part of the exclusion zone.

Wildlife especially likes the empty human settlements, which provide a host of unusual and valuable hidey holes. 'Abandoned buildings attracted many species of wild animals that use them to rest and bring up their young', says a report from the radiation ecological reserve.[34] Badgers burrow in cellars, barns and under the concrete slabs of roads. Wild boar rest in sheds. Roe deer, foxes and pine martens feed in the overgrown gardens. Elk meet in the abandoned villages. Owls and kestrels nest in empty buildings. The reserve has the country's largest lynx population. Bears have arrived.

Chernobyl is not quite alone in its radioactivity. Testing sites for nuclear bombs have suffered much more. Half a century ago, the waters of the Bikini Atoll in the Northern Pacific were boiled by 23 US nuclear weapons tests over twelve years. The 15-megaton Bravo test destroyed three islands, irradiated the ocean and blasted millions of tonnes of coral, sediment and marine life into the air. Yet today two-thirds of the atoll's former coral species are back, along with some newcomers.[35] Nobody would want more such places, but their ecological survival still tells us something important about nature's powers of recovery – something that should help the world undo some of the environmental devastation of the 20th century.

*

Conservationists need to take a hard look at themselves and their priorities. They must learn from Puerto Rico and Chernobyl, the Tilbury ash heap, the Bikini Atoll, the feral streets of Chicago and the wider world of novel ecosystems. Nature no longer congregates only where we expect to find it, in the countryside or in 'pristine' habitats. It is increasingly eschewing formally protected areas and heading for the badlands. Nature doesn't care about conservationists' artificial divide between urban and rural, or between native and alien species. If conservationists are going to make the most of the opportunities in the 21st century to help nature's recovery, they must put aside their old certainties and ditch their obsessions with lost causes, discredited theories and mythical pristine ecosystems.

One of the few conservationists I have met who is willing to make the change is Peter Kareiva, chief scientist of The Nature Conservancy in the US, the world's largest and richest environmental organization. Ironically, his members are among those nature enthusiasts who are most wedded to the old ecology – the most reluctant, as Kareiva puts it, 'to shed the old paradigms'. TNC's slogan used to be 'Saving the last great places on Earth'. But Kareiva rejects the whole idea. 'Conservation's continuing focus upon preserving islands of [old] ecosystems in the age of the Anthropocene is both anachronistic and counterproductive', he argued in a polemic for the California-based Breakthrough Institute.[36]

There are hard choices to be made. Conservationists often suggest that protecting each last individual native species is somehow essential to maintaining the 'ecological services' that nature provides for us – services such as carbon storage and maintaining the chemistry of the oceans; protecting watersheds and maintaining river flows; pollinating plants and dispersing seeds; maintaining soils and preventing runaway erosion. But that argument is a romantic illusion. Those services are best done by the species on hand that do it best. In much of the world that increasingly means nature's pesky, pushy invaders.

Conservationists have 'grossly overstated the fragility of nature, arguing that once an ecosystem is altered, it is gone forever', Kareiva says.[37] The trouble is that the data simply do not support the idea. Conservation scientists spend too little time investigating how ecosystems change when invaders come in or humans disrupt their operation. A narrow pursuit of evidence of 'harm', driven by invasion biologists, has blinkered researchers. And so has their pervasive belief that stability is the norm and change somehow abnormal. Neither is true. Nature is rarely in a steady state. It is the dynamics that matter, and for too long researchers have denied this.

By its own measures, conservation is failing, Kareiva says. Many protected areas in which conservationists have invested so much are about as true to nature as Disneyland. From the Serengeti to Yellowstone, from the Amazon jungle to Siberia's Pleistocene Park, these are managed ecosystems. Conservation cannot promise a return to pristine, pre-human landscapes. It needs to 'jettison idealized notions of nature, parks and wilderness – ideas that have never been supported by good conservation science – and forge a more optimistic, human-friendly vision'.[38]

As a leader in *New Scientist* magazine in May 2013 argued, conventional conservation has focused 'on two goals: saving threatened species and restoring Earth to how it used to be. Both are doomed. The first misses the point, and the second is impossible ... The concept of natural has outlived its usefulness in conservation.' Ecosystems, it said, are not worth preserving 'in aspic, or rewinding to some romantic, pristine past'.[39]

I would add one more thing. The more damage that humans do to nature – through climate change, pollution, grabbing land for intensive agriculture and plantation forestry – the more important alien species and novel ecosystems will be to ensuring nature's survival. Aliens are rapidly changing from being part of the problem to part of the solution. And in a world where supposedly pristine habitats require constant micro-management to keep them going, where they are increasingly like theme parks for conservation scientists, the truly wild lies elsewhere. It is in the unmanaged badlands and novel ecosystems. The bits of nature we don't cosset and pamper. The new wild.

12
Call of the New Wild

I first came across ecologist Chris Thomas a decade ago. He had a paper that was taking climate change science by storm. It warned that global warming would wipe out a quarter of the world's species. Just two degrees of warming by 2050 would see millions of species, if not gone, then, as he and eighteen co-authors put it, 'committed to extinction'. It was the gloomiest prediction yet on the fate that awaited nature as we turned up the planet's heat.[1]

Thomas's team had analysed the climate zones within which species are currently found, figured out from climate model predictions how fast the zones would move, and made estimates about which species could keep up. The finding that only three-quarters could do so made it as a headlined finding of the next report from the UN's Intergovernmental Panel on Climate Change (IPCC). The paper has been hugely influential, with Google Scholar finding 3,300 citations – more than almost any single paper on climate change.

Thomas was so alarmed by his findings that he and his colleagues began trials to physically move species to help them keep up. They took 500 marbled white butterflies (*Melanargia galathea*) from York in northern England, the far north of their current range, and released them from the back of a car about 100 kilometres further north at a quarry in Durham.[2] The marbled whites are doing well. But not everyone likes interfering with nature in this way. Daniel Simberloff, the guru of invasion biologists, called this forced migration 'tantamount to ecological roulette'. He noted that it 'signals the emergence among some conservationists of a new philosophy ... at odds with the traditional objective of preservation'. His fear was that 'even species that are threatened in their native ranges could become invasive in a new evolutionary context'.[3] Even in the face of climate change, species should be kept where they belong.

Thomas counters that there are equal, and perhaps greater, risks from inaction. He thinks that in an era of rapid climate change, organized translocations of species will be 'the only realistic way to maintain wild populations of some species'.[4] If we are messing with the climate, we should at least try to help nature keep up.

Thomas's initial assessment of the ecological meltdown likely to accompany climate change is surely far too gloomy. It rests on the idea that species occupy their strict climatic and ecological zones and cannot prosper if that environment changes. Many species may turn out to be much more versatile and pushy. They may, where necessary, find their ways to new environments – without waiting for a lift in Thomas's car.

Climate change is certain to be a major driver of species migrations, whether Simberloff welcomes them or not. In Britain there is already a trickle of arrivals crossing the English Channel from France and then pushing north. Taking advantage of warmer climes, the little egret (*Egretta garzetta*) first showed up in Britain in 1996, and is now widespread in British wetlands. Should Brits be sending it home? Some conservationists take Simberloff's zero-tolerance approach. But the bird conservationists at the Royal Society for the Protection of Birds are welcoming. And if we should welcome the little egret, then why not give a helping hand to other species in search of a haven?

Thomas has proposed Britain as a sanctuary for species at risk elsewhere. After all, it has no undisturbed terrain and no globally endangered species to lose. It would be an ideal refuge for lost species. The Iberian lynx, the most endangered cat in the world, would love hunting British rabbits. Britain once had lynx, and its rabbits originally came from Iberia. So why not?[5]

*

When I visited Thomas in York in late 2013, however, he had a much more optimistic story to tell – about nature's ability to bounce back from human threats. He was, he said, now convinced that human activity, including climate change, could trigger an evolutionary explosion that might counterbalance the extinctions, and could leave us with more species than before. And alien species were a big reason behind his new optimism. They were the dynamic element, the survivors, the gene-spreaders and all-round colonists that could turn ecological disaster to evolutionary triumph. The thinking was simple. Extinctions would create the opportunity for evolution to go into overdrive. Every species lost would create space for new species to move in. And the most invasive of the aliens would be the ones best able to take advantage. The species that conservationists most fear are precisely the ones that nature most needs.

Thomas had just published a short note in the science journal *Nature* entitled 'The Anthropocene Could Raise Biological Diversity'. In it, he said that 'it is time to rethink our irrational dislike of invading species'.[6] He told me: 'We worry about the extinction of species in the era of humans. We are right to. But the seeds of recovery are already visible. New species are beginning to emerge. Of course many of the new species will fail. But others will become the new lineages of the future.' Nature as a dynamic force will be reborn.

This is controversial stuff. The conventional narrative is that climate change may be leading the world towards what some term the Earth's sixth great extinction. One that is potentially as dramatic as what happened when an asteroid hit the Earth 65 million years ago and wiped out probably 50 per cent of species, including the dinosaurs. Thomas's own research a decade before did much to make the case for climate change as the prime cause of that extinction. Others stress the role of habitat loss and the spread of alien species. The argument about aliens is that by extinguishing local endemics and replacing them with global generalists they will reduce the world's overall stock of species. They will cause the 'homogenization' of nature. Niche players will be lost and the weedy generalists will spread across the land.

Several questions are raised here. One is whether global biodiversity actually matters to nature. The other is whether aliens will destroy it by accelerating extinctions or help nature recover from it by stimulating evolutionary forces.

On the first question, there is actually no compelling evidence that the planet's total stock of species is of any intrinsic importance to the functioning of nature. We know that local ecosystems may be more resilient to change if they contain more species. That, incidentally, is another reason why alien species that add to local biodiversity ought to be smiled on. But there is no research that I know of pointing to such a value from global biodiversity. This suggests to me that our current preoccupation with how much biodiversity the planet as a whole is losing may be rather ill-founded. Local biodiversity matters much more. Species act locally; we should think locally about them.

On the second question, Thomas points out that all five earlier great extinctions led to a burst of evolutionary renewal. Yes, there was a short-term loss of species, but even as it happened, nature was gearing up to make good the loss. The Darwinian engine of evolution raised its game. Thus, the demise of dinosaurs created space for mammals to evolve. So, he argues, why not look forward to the new diversity of the Anthropocene?

Already, new hybrids and species are forming as native and alien species both adapt to changing conditions, and as their genes mingle. The new science of molecular genetics is revealing these evolutionary stirrings. According to Thomas: 'There is an extraordinary amount of hybridization going on. Genes are jumping around. Darwin talked about a tree of life, with species branching out and separating. But we are discovering it is more like a network, with genes moving between close branches of the evolutionary tree, as related species interbreed. This hybridization quickly opens up new evolutionary opportunities.' The argument is that the homogenization of global biodiversity will be counteracted by new evolution.

Alien species are a big part of Thomas's evolution revolution, turning up in unexpected places and churning the genetic soup. 'In Britain, hybridization involving introduced plant species seems to be happening at least as fast as native species are going extinct', he said. Thomas is not alone in his thinking. 'Evolution can be very rapid under the right conditions. Climate change is going to be one of those things where the conditions are met', says Arthur Weis of the University of Toronto.[7]

As Thomas and I looked out of his lab window on the bleak winter campus of York University, none of this seemed likely. Nature looked somnolent. But his computer screen was soon flashing with papers from across the world describing the unexpected extent of promiscuous genetic exchange. Tim Benton of Leeds University, who had taken over from Thomas when he moved to York, found major genetically transmitted changes in wild soil mites within fifteen generations of them changing their environment. The genetic changes doubled the age at which the mites reached adulthood.[8]

That was in the lab. But there are plenty of stories from the field. Oxford ragwort (*Senecio squalidus*) originally lived on the slopes of Mount Etna in Sicily. It was brought to Britain in about 1700 by botanist William Sherard, and was tended in the Oxford Botanic Garden. Eventually it escaped, first working its way along the walls of Oxford colleges where, by the late 18th century, it was reportedly 'very plentiful'.[9] Later, as Britain developed a network of railway lines, it adopted the stone beds of the tracks as a new habitat comfortingly similar to the lava beds back home. It made its way along the railway to London and elsewhere. At some point on this journey, it mixed its genes with *Senecio* cousins such as the sticky groundsel (*Senecio viscosus*), creating a number of hybrids that eventually turned into fully fledged new species. Among them are *Senecio cambrensis* and the most recent addition *Senecio eboracensis*, which was recently named York groundsel, after the city it was found in.[10] Thus a species introduced to Britain under circumstances that would horrify people defending native biodiversity has become a dramatic example of evolution in action. The alien ragwort is not destroying biodiversity; it is creating it.

The Oxford ragwort is far from alone. A new fertile hybrid primrose, *Primula kewensis*, arose when *Primula verticillata* from Africa met *Primula floribunda* from the Himalayas at Kew Gardens in London. And whenever the southern European *Rhododendron ponticum* meets its Appalachian cousin, *Rhododendron catawbiense*, the result seems to be a new hybrid. One now widely found in north-east Scotland is both fertile and frost-resistant.

Often, migrants evolve in the course of adapting to their new surroundings. That has happened to the thistle-like plants called *Centaurea melitensis* that missionaries took from Spain to California in the late 18th century. Already, the migrants have developed 'reproductive incompatibility' with their cousins back home. They are, in effect, a new species. Similarly, European house sparrows have grown apart from their forebears since arriving in North America in the mid-19th century.[11] And sockeye salmon in Washington State turned into a new species within 56 years, or thirteen generations, of their introduction to new lakes. Nature doesn't hang about.[12]

Native species evolve too, so they can take advantage of new arrivals. A new hybrid of the *Rhagoletis* fruit fly in North America has developed the ability to colonize invasive honeysuckles, while its parent species continues to feed on native plants.[13] All this will happen more and more often as a result of alien introductions, says Thomas. 'Evolutionary origination is accelerating, as populations and species evolve, diverge, hybridize and speciate in new surroundings', Thomas wrote in the unpublished original transcript of his *Nature* paper.

Those who fear alien species also often fear hybridization. Simberloff calls genetic mingling between natives and aliens 'a sort of genetic extinction'. He says that 'hybrid swarms' will take out the pure natives.[14] He cites the case of Canada's ruddy duck's cuckolding with Europe's white-headed duck, which we looked at in Chapter 6. Another example is Scottish wildcats interbreeding with feral domestic cats. Such swamping can happen, agreed Thomas. 'But new genes from alien species usually only invade the genome of an existing species if they confer some advantage. They often help them thrive.' This is nothing more than Darwinian survival of the fittest, he argues. We should look at what is gained, not what is lost.

Hybridization can turn minnows into monsters, but it can also create new species, new diversity and new virtues. If we want to encourage evolution, we must embrace successful alien species. In a world of fragmented and abused nature, with many species under threat, nature needs them.

Thomas accepts that we will sometimes recoil at the kind of dynamic and unpredictable change he is talking about. 'Things are happening so fast that we see ecological transformations in our individual lifetimes. It is human nature to be worried about that. I sometimes pull up alien weeds that I see in the fields around where I live. But that is an emotional response. Intellectually, I see nothing wrong with most of them.' We should, he said, 'not confuse change with damage, or think of alien species as bad and natives as good'. Nature is always changing and adapting. Species are going to have to move to survive, whether in the face of climate change, other human activity, or simply the ongoing change that constantly marks out nature. 'A narrow preservationist agenda will reduce rather than increase the capacity of nature to respond to the environmental changes that we are inflicting on the world', he said. 'We need to think less about keeping things just the way they were – because it is impossible – and more about promoting the new.'[15]

*

This amounts to the beginning of a manifesto for the new wild. It has many adherents. Christoph Kueffer of the Swiss Federal Institute of Technology in Zurich says novel ecosystems 'represent the wild lands of the future, the self-organized response of nature to anthropogenic impacts'.[16] 'There is no point in debating what is natural and what is novel. It is all novel', he says. The traditional idea that biodiversity was 'conserved most effectively by protecting nature from human influence does not work any longer. Humans and their impacts are omnipresent; a new paradigm is needed for guiding conservation action.'

Conservationists invest their time and money to fight what Kueffer calls 'the ghosts of past invasions, always one step behind'. In doing so, says Kueffer, they 'increasingly run the risk of preventing the introduction of species that would be useful in the new circumstances'. Martin Schlaepfer of the State University of New York agrees. 'Any conservation strategy that eradicates species simply because they are non-native could undermine the very biological entities that may be the most likely to succeed in a rapidly changing world.'[17]

Ecosystems, says Thomas, 'don't have a pre-ordained cast list. They dissolve and create entirely new communities', like turning the mirrors in a kaleidoscope. Novelty is the norm in nature. It always has been. Ecologists touring the world thousands of years ago would have seen most of the same species we see today. But they would have been arranged differently. At the end of the last ice age – the last time we had big climate change – tree species migrated north as the ice retreated. They were reclaiming old territory, but they didn't go back at the same speed, or to the same places. As a result, new combinations of species emerged. New ecosystems. In New England, beech and hemlock trees are today regarded as long-standing ecosystem companions. Yet pollen studies show that as the ice retreated, hemlock appeared in Massachusetts 9,000 years ago, while it was another 2,000 years before beech showed up.[18]

This reconstruction was made by ecologists in the 1980s. Thomas says that it was one of the pieces of research that first opened the eyes of his generation to the idea that the Clementsian orthodoxy might be wrong. 'Back then, ideas about ecological communities and co-evolution were dominant. So when ecologists started reporting this stuff, I remember thinking: wow. It contradicted everything I had been taught. It started a much more individualistic view of things. We realized that botanical communities don't move *en masse*. Individual organisms move. In the end, it is they that are the biological building blocks, not ecosystems.'

Another young researcher who took that message was Stephen Jackson of the University of Wyoming. Few things in nature stay stable for long, he concluded. 'Change, including rapid and disruptive change, is natural', he says. The world 'has been in continual environmental and ecological flux throughout its history. Ecological novelty is widely perceived as a threat to conservation. However, it also comprises a reality and, more importantly, an opportunity.'[19] It is also a necessity. The old wild will have to be constantly micro-engineered to keep it functioning in an increasingly unfamiliar world. The rest of nature – an increasingly large proportion of it – will be living in 'novel' ecosystems, where alien species live cheek-by-jowl with natives, all fitting in as best they can. They will have their mishaps and disasters, their grand advances and ignominious retreats. But they will exhibit the abiding traits of real nature – transience, dynamism and contingency, rather than stability, permanence or predictability.

Here we face a central paradox of conservation in the 21st century and beyond. Traditional wild lands – the old-growth forests and other historic habitats – will in future be the places most dependent on human intervention for their survival. In a world of climate change, where the old wild is hemmed in by human activity, these ecosystem islands will increasingly resemble museum pieces, time capsules and experimental labs for scientists. They will not be 'wild' in any true sense. On the other hand, the novel ecosystems, the make-do-and-mend places, will be the ones able to stand on their own two feet. They will be the new wild.

Nobody wants species to go extinct. Perhaps the pace of loss can be slowed. I hope so. But it is happening. What should increasingly matter for conservationists is finding ways to help nature regenerate. That should mean supporting the new, rather than always spending time and money in a doomed attempt to preserve the old. Embracing the new – striving to find what works in the new world – will often mean making peace with old enemies, because successful alien invaders are increasingly what works in nature. They are the winners and likely saviours. They are the ones able to reclaim the abandoned fields, to rise up through the cracks of urban concrete jungles, to kick-start the revival of degraded forests and shrug off the rigours of climate change. As the late Stephen M. Meyer put it, conventional notions of wild lands 'are just fantasies now. There is virtually no place left on Earth that fits this definition.'[20]

Sometimes the colonists will not be the species that, aesthetically speaking, we would have chosen. But that is our doing. We are, wrote Meyer, 'making the planet especially hospitable for the weedy species … that thrive in continually disturbed human-dominated environments'. He meant no disrespect in calling them weeds. But the world's weeds – the scrappers and generalists, the versatile and opportunist, the cockroaches and coyotes, the rats and rhododendrons – seem set to inherit the Earth.

We have to embrace them. We must, along the way, give up some of our romantic ideas about nature being passive and fragile. We should take heart from the growing realization that nature is actually dynamic and can-do. This should change our approach to nature conservation. In fact, conservation as currently practised becomes the enemy. By seeking only to conserve and protect the endangered and the weak, it becomes a brake on evolution and a douser of adaptation. If we want to assist nature to regenerate, we need to promote change, rather than hold it back.

In this rebooted vision of nature, species will move around much more than in the past, whether by deliberate human hand, their own volition, or by accident. Those species that are opportunistic, versatile, aggressive and prolific will globalize their activities. In the title of a rather ignored book a few years ago, journalist Garry Hamilton called these colonists and vagabonds the 'super species'. The house sparrow and Argentine ant, the water hyacinth and wild boar, the kudzu and cat, the rat and zebra mussel have shown the way.[21] Their success represents not nature's degradation but its fight-back against us. They are the ones that move in when we screw things up. The super species will take charge as what Meyer called the 'relic species' either perish or are confined to conservation's intensive care wards.

*

All that said, we humans will want to intervene, to preserve what we like and need for our own ends. We will sometimes need to defend against inconvenient invaders of our spaces, destroyers of our most cherished natural companions, pests and diseases. And we will continue to have our favourite species. Many, like the giant panda, Sumatran rhino and California condor, already survive only thanks to our constant protection. There is no harm in that. These living trophies have meaning for us. But we should not kid ourselves that they are the future of nature. In preserving them we will be serving our own desires, not nature's needs. Nature is travelling in a different direction.

The new fad among conservationists is rewilding, the recreation of large areas of land for nature. Why not, if we can? But what kind of wild do we hope to create in this way? For many, the ambition is the recreation of ancient 'pristine' habitats. They want to bring back totemic 'native' species, like wolves and bears in the Scottish highlands and bison herds across the American Midwest. They want to place them in habitats as close as can be managed to the pristine. This again is a human cultural choice. In the Americas, it often means reproducing the landscapes before Europeans showed up – 1491, in other words. In Australia it is similar, though the date is rather later. Neither is in any sense pristine. As we have seen, both landscapes were transformed by their previous occupants.

In Europe, the ambition is to go further back. But how far? Britain probably lost its extensive wildwoods 5,000 years ago, as hunter-gatherers gave way to Neolithic farmers.[22] Many bogs much loved by environmentalists turn out to have been the result of early farming activity. Go back further and most of Britain was covered in ice. Before that, during the previous interglacial, it had hippopotamus, lions, hyenas and Barbary apes.[23]

In reality, all efforts at rewilding are compromises, both with the reality of the wider landscape and with neighbours. Fairly typical is the Scottish highland estate known as the Alladale Wilderness Reserve, a 9,000-hectare glen bought by Paul Lister, the heir to a flat-pack furniture fortune. He wants to recreate the natural pine forests that once covered most of the area, and to populate them with some of their ancient fauna. His wish list includes elk that disappeared around 3,000 years ago, brown bears not seen for 2,000 years, lynx that stuck around until 1,500 years ago, and wolves that survived until just 300 years ago. But the irony is that to do this requires yet more intervention.[24]

So far, Lister's rewilding has been slow, and doesn't feel so very wild. He has had to put up fences to protect the million or so saplings he has planted from the large numbers of native red deer that wander the highlands – natives left over from the days of the old forests. He briefly imported Swedish elk and some bison, but 'we couldn't keep them in a serious natural state. They looked good in a TV show filmed here, but they were a zoo item and we got rid of them after that.' He has a small enclosure housing a trio of wildcats. They are fed with rabbit and fish like the mangy tabbies they resemble but, to his disappointment, refuse to breed. Russian oligarchs come up here to hear the rutting stags bellowing across the glen. 'They bring their own armed security guards', my taxi driver told me. Fearing assassins rather than wolves. While walking the peat moor above the glen, I plugged my ears as three Tornado fighter aircraft shot past us, following the contours of the valley below.

'We are not recreating the past here', Lister admitted that evening in the Alladale Lodge, an imposing Victorian stone hunting lodge built for an Indian maharaja. 'This is a man-made landscape and will remain so. If we could introduce a couple of packs of wolves, and maybe a dozen brown bears, that would be my dream. I'd die happy.' In truth he is creating a bizarre hybrid novel ecosystem. That's fine, but it's not terribly wild.

Most other efforts similarly suggest more novelty than wilderness. The Dutch decided to give nature the 5,600-hectare Oostvaardersplassen polder that had been reclaimed from the North Sea in 1968. A literal interpretation of rewilding would see the dykes breached and the land returned to the sea. But instead the scientist in charge of the government's nature reserves, Frans Vera, decided to try to recreate a slab of European river delta from before the days of dykes and windmills. He planted bulrushes and reed beds as havens for wetland birds like bitterns, spoonbills, marsh harriers and white-tailed eagles. More interestingly, he also brought in Heck cattle – the result of a controversial attempt by German scientist Heinz Heck in the 1930s to breed back extinct wild cattle known as aurochs – and Polish konik horses, which some believe to be close relatives of Europe's extinct wild horses. Less sexy, but arguably more native, are the red deer.

It turned out that most Dutch citizens do not want too much wildness. When the animals started dying from hunger there were widespread protests. The scientists defended the deaths, saying that would be the normal wild way of things. But one animal-rights group sued. The group lost, but now the reserve sends in people with guns to shoot starving animals, to reduce their suffering. As many as a fifth of the reserve's animals end their days this way.[25] Again, the boundary between rewilding and a large zoo seems blurred.

The North American version is the return of bison. I went to northern Montana, where the American Prairie Foundation, a land trust set up by the environment group WWF, is buying cattle ranches south of the high plains railroad city of Malta and putting in bison. The project is bankrolled by wealthy environmentalists, including members of the Mars family. It owned some 50,000 hectares at the time of my visit, and had access to much more public land. This is just part of a wider ambition to return to the days when 70 million bison roamed the northern Great Plains, and wolves, cougars, grizzlies, bighorn sheep, elk and prairie dogs were not uncommon – the recreation of what *National Geographic* calls 'an American Serengeti'.[26]

Private herds are emerging on private land through the state. Media magnate Ted Turner, one of the largest landowners in the US with around half a million hectares, gets his bison from surplus animals at Yellowstone National Park. But the animals are only half-wild. Most contain many genes from domestic cattle. And the land on which they graze is fenced. When I asked to visit Turner's spread, I was refused. This is privatized nature, and the creatures are, in effect, farmed. Turner has opened a chain of restaurants in which to sell his 'bison burgers'. That's fine. I don't object. But it is not wilding; it is taming.[27]

At its most ambitious, rewilding extends to bringing back extinct megafauna. *Jurassic Park* may be science fiction, but the idea of a Pleistocene Park is real. There is already a park with that name way up in the sub-Arctic steppes of north-eastern Siberia, not so far from the Bering Straits. It is the creation of Russian scientist Sergey Zimov, director of the Northeast Science Station in Cherskiy. The park already contains musk-oxen and Yakutian horses, but he would dearly like someone to genetically recreate woolly mammoths, which existed there until some 10,000 years ago. It may be possible, by harvesting fragments of DNA from museum relics and fleshy fragments of mammoth occasionally unearthed from melting Siberian permafrost. The trick would be to reassemble the DNA sequences and implant them in the animal's modern-day cousin, the Asian elephant.[28]

The Californian resurrection guru Stewart Brand – former hippy hero of the Whole Earth Catalogue and now proprietor of the Long Now Foundation – proposes the mammoth for revival, along with other more recently extinct species like the great auk, quagga, thylacine and Steller's sea cow. Brand argues that, in this way, 'the conservation story could shift from negative to positive, from constant whining and guilt-tripping to high fives and new excitement'. I have some sympathy with that. I agree too with Brand that 'nature is not broken, nor is it particularly fragile'. But his project feels more like creating a very expensive novel ecosystem than rewilding.[29]

Whatever the hopes of their founders, these engineered landscapes will only keep going in a rapidly changing world if they are constantly tended, like outsize gardens. Whatever they become, they will not be wild. And why the constant tug backwards? The new wild will not be the same as the old wild. Let's stop pretending otherwise.

*

The great American ecologist Edward O. Wilson predicts that

restoration ecology will become a major activity for conservationists in the 21st century. It already is. Rivers are being restored, along with forests and wetlands and natural grasslands. Some scientists are outright in their condemnation of this. Arthur Shapiro of UC Davis, a butterfly expert, is among them. He wrote an open letter in 2011 to the San Francisco Planning Department complaining about its plan to set up natural areas in the city. 'I am frankly appalled that San Francisco is considering major expenditure directed towards "restoration ecology" … a euphemism for a kind of gardening informed by an almost cultish veneration of the "native" and abhorrence of the naturalized, which is commonly characterized as "invasive".'

Shapiro accused the planners of trying to recreate 'a simulacrum of what is believed to have been present at some (essentially arbitrary) point in the past'. The result, he said, will 'almost never be capable of being self-sustaining without constant maintenance. The context has changed, the climate has changed, the pool of potential colonizing species has changed, often drastically.' This kind of rewilding, in other words, won't be wild at all. Existing ecosystems, he said, 'are freeze-frames from a very long movie. Ecological change is the norm, not the exception. The ideology that informs restoration ecology basically seeks to deny evolution and prohibit change.'[30] It would, he said, be a 'Sisyphean' task – after the mythical king of Corinth doomed to eternally roll a rock up a hill, only for it to fall back to the bottom every time he neared the summit.

Many ecological restorers and conservation scientists think this criticism is a gross caricature of what they intend. 'Restoration ecologists do not aim to recreate the past, but rather to re-establish the historical trajectory of an ecosystem before it was deflected by human activity', says Simberloff.[31] But it is hard to see how that happens without massive and continued environmental engineering. Even a world suddenly without humans would be hugely changed. And we are stuck with one degree of global warming even if we turned off every fossil fuel burner tomorrow. I don't agree with Shapiro, even though I am sure much of what he says is right. The people of California have the perfect right to attempt the creation of a *Lost World* of Sir Arthur Conan Doyle's imagining, complete with dinosaurs if they can find the DNA. But it will require constant tending. It won't be natural. And it won't be wild.

Rewilding efforts meet a human desire to see more nature on a larger scale than we have been used to. But if such efforts are to become something other than a large zoo or a theme park for scientists, if they are to be nature 'red in tooth and claw', open to evolutionary change and able to contribute to genuine ecological revival, then we have to let go and let nature take its course, however novel, however divorced from what we might like to think of as pristine nature.

More interesting to me than a big zoo on a polder – or even refilling the Great Plains with herds of bison – is the return of wolves across Europe. They are arriving under their own steam and in their own time. And they are not 'returning' to traditional habitats or anything created especially for them. Now largely safe from hunters, wolves are mooching back onto abandoned farmland and into the suburbs. This, to me, is genuine rewilding. The new wild.

There are now an estimated 12,000 wolves migrating from old refuges in Eastern Europe as far west as France and Spain. Joining them are lynx, spreading out from the Carpathians and Balkans, jackals, brown bears, wolverines, Alpine ibex and European bison. Hundreds of thousands of beavers are gnawing away at riverbank trees. Millions of wild boar are everywhere. These 'returns' are mostly into inhabited landscapes. And that is the way it will mostly be. In the Anthropocene, fortress conservation is a doomed enterprise. Humans are an inescapable part of the landscape. There are no pristine ecosystems and no blueprints for what they might be. Any vision of the pristine past that we choose will require constant tending.

The old wild is dead. But the new wild is flourishing, and will do better if we allow it to have its head. It is there in hybridizing rhododendrons; in rare bees and spiders appearing amid the badlands of the Thames estuary; on Ascension Island's Green Mountain; in Chernobyl's exclusion zone; across the bushlands of tropical Africa and the regrowing rainforests of Borneo; along the creeks of San Francisco Bay and the lava flows of Hawaii; in the cinnamon forests of the Seychelles; among the advancing wolves of suburban Europe and the birds flocking to keep up with climate change; on Surtsey and Krakatoa; among the *Caulerpa* beds of the Riviera; in the Great Lakes; and amid the call of the *coqui* across the abandoned plantations of Puerto Rico. Nature never goes back; it always moves on. Alien species, the vagabonds, are the pioneers and colonists in this constant renewal. Their invasions will not always be convenient for us, but nature will rewild in its own way.

That is the new wild.

Appendix

Latin names

I have included in the text the Latin names of those species that are most important to the story. But many other species are mentioned in passing, using their common names. Some people like to have these names, and they do bring clarity where there may be confusion. So here is a list of common names used in the text, followed by their Latin names. It is not exhaustive. I left out many domesticated crops, animals and harvested stocks like fish, as well as other plants, animals and insects where I am talking about whole genera or families of species, or where the species are known only by their Latin name.

African land snail: *Achatina fulica*
African tulip tree: *Spathodea campanulata*
Aldabra giant tortoise: *Geochelone gigantea*
Algerian sea lavender: *Limonium ramosissimum*
Alpine ibex: *Capra ibex*
American bullfrog: *Lithobates catesbeianus*
American mink: *Neovison vison*
American northern mockingbird: *Mimus polyglottos*
American willowherb: *Epilobium ciliatum*
Amur clam: *Corbula amurensis*
Arctic tern: *Sterna paradisaea*
Argentine ant: *Linepithema humile*
Argentine cactus moth: *Cactoblastis cactorum*
armadillo: *Dasypodidae*

Ascension crake: *Mundia elpenor*
Ascension frigatebird: *Fregata aquila*
Ascension Island parsley fern: *Anogramma ascensionis*
Ascension lily: *Hippeastrum reginae*
Ascension rail: *Atlantisia elpenor*
Ascension spurge: *Euphorbia origanoides*
Asian carp: *Catla catla*
Asian kelp: *Undaria pinnatifida*
Asian red scale insect: *Aonidiella aurantii*
Atlantic petrel: *Pterodroma incerta*
Australian paperbark tree: *Melaleuca quinquenervia*

badger, European: *Meles meles*
barn owl: *Tyto alba*
bean plataspid: *Megacopta cribraria*
Bermuda cedar: *Juniperus bermudiana*
bilby: *Macrotis lagotis*
bird's-foot trefoil: *Lotus corniculatus*
bison, North American: *Bison bison*
bison, European: *Bison bonasus*
blackbird: *Turdus merula*
black grouse: *Tetrao tetrix*
black guillemot: *Cepphus grylle*
black kite: *Milvus migrans*
black rat: *Rattus rattus*
black stork: *Ciconia nigra*
blacksnake: *Pseudechis*
blackthorn: *Prunus spinosa*
black wattle: *Acacia mearnsii*
blue and gold macaw: *Ara ararauna*
blue fleabane: *Erigeron acris*
bog rosemary: *Andromeda polifolia*
boll weevil: *Anthonomus grandis*
bracken: *Pteridium aquilinum*

Brahminy blind snake: *Ramphotyphlops braminus*
bramble: *Rubus fruticosus*
Brazilian pepper-tree: *Schinus terebinthifolius*
Brazil nut: *Bertholletia excelsa*
breadfruit: *Artocarpus altilis*
brown bear: *Ursus arctos*
brown rat: *Rattus norvegicus*
brown skua: *Stercorarius antarcticus*
brown tree snake: *Boiga irregularis*
brushtail possum: *Trichosurus vulpecula*
buddleia: *Buddleia davidii*
Burmese python: *Python molurus bivittatus*

California cordgrass: *Spartina foliosa*
California clapper rail: *Rallus longirostris obsoletus*
California condor: *Gymnogyps californianus*
California quail: *Callipepla californica*
Canada goose: *Branta canadensis*
candlenut tree: *Aleurites moluccana*
cane toad: *Rhinella marina*
Canvey Island ground beetle: *Scybalicus oblongiusculus*
cat, domestic: *Felis catus*
cattle egret: *Bubulcus ibis*
cheatgrass: *Bromus tectorum*
Chinese mitten crab: *Eriocheir sinensis*
Chinese water deer: *Hydropotes inermis*
Christmas Island red crab: *Gecarcoidea natalis*
cinnamon: *Cinnamomum verum*
cochineal: *Dactylopius coccus*
cocksfoot: *Dactylis glomerata*
cogon: *Imperata cylindrica*
Colorado beetle: *Leptinotarsa decemlineata*
common frog: *Rana temporaria*
coqui: *Eleutherodactylus coqui*

cottonwood: *Populus deltoides*
couch grass: *Elymus repens*
coyote: *Canis latrans*
coypu: *Myocastor coypus*
Cuban tree frog: *Osteopilus septentrionalis*

dandelion: *Taraxacum officinale*
devil's guts: *Aegopodium podagraria*
dingo: *canis lupus dingo*
distinguished jumping spider: *Sitticus distinguendus*
dog, domestic: *Canis lupus familiaris*
dog rose: *Rosa canina*
donkey: *Equus africanus asinus*
duck-billed platypus: *Ornithorhynchus anatinus*

eastern mudsnail: *Ilyanassa obsoleta*
edible dormouse: *Glis glis*
Egyptian goose: *Alopochen aegyptiaca*
eland: *Taurotragus oryx*
elk: *Cervus canadensis*
emerald ash borer: *Agrilus planipennis*
Eurasian beaver: *Castor fiber*
Eurasian spoonbill: *Platalea leucorodia*
Eurasian water milfoil: *Myriophyllum spicatum*
European honeybee: *Apis mellifera*
European rabbit: *Oryctolagus cuniculus*
European starling: *Sturnus vulgaris*

fallow deer: *Dama dama*
ferret: *Mustela putorius furo*
fire tree: *Morella faya*
Florida panther: *Puma concolor coryi*
French broom: *Teline monspessulana*
freshwater crocodile: *Crocodylus johnsoni*

Galapagos dove: *Zenaida galapagoensis*
Galapagos rail: *Laterallus spilonotus*
gallant soldier: *Galinsoga parviflora*
Gambian pouched rat: *Cricetomys gambianus*
giant hogweed: *Heracleum mantegazzianum*
giant panda: *Ailuropoda melanoleuca*
ginger: *Zingiber officinale*
glow-worm: *Phosphaenus hemipterus*
goat: *Capra aegagrus hircus*
golden-headed lion tamarin: *Leontopithecus chrysomelas*
Gough bunting: *Rowettia goughensis*
Gough moorhen: *Gallinula comeri*
great crested newt: *Triturus cristatus*
great white egret: *Ardea alba*
green iguana: *Iguana iguana*
green turtle: *Chelonia mydas*
grey-headed flying fox: *Pteropus poliocephalus*
grey squirrel: *Sciurus carolinensis*
grizzly bear: *Ursus arctos ssp.*
Guanacaste tree: *Enterolobium cyclocarpum*
guava: *Psidium guajava*
gypsy moth: *Lymantria dispar dispar*

hairy-stemmed Canadian fleabane: *Conyza canadensis*
Hawaiian rail: *Porzana sandwichensis*
hedgehog: *Erinaceus europaeus*
hemlock: *Conium maculatum*
Henderson petrel: *Pterodroma atrata*
herring gull: *Larus argentatus*
Himalayan balsam: *Impatiens glandulifera*
hoary cress: *Lepidium draba*
horse chestnut: *Aesculus hippocastanum*
Hottentot fig: *Carpobrotus edulis*
house mouse: *Mus musculus*

house sparrow: *Passer domesticus*

Iberian lynx: *Lynx pardinus*
ivy-leaved toadflax: *Cymbalaria muralis*

Jamaican fruit bat: *Artibeus jamaicensis*
Japanese cherry: *Prunus serrulata*
Japanese dodder: *Cuscuta japonica*
Japanese goby: *Tridentiger trigonocephalus*
Japanese knotweed: *Fallopia japonica*
Japanese oyster: *Crassostrea gigas*
Japanese white-eye: *Zosterops japonicus*

kangaroo rat: *Dipodymus ordii*
kelp gull: *Larus dominicanus*
kilka: *Clupeonella delicatula*
koala: *Phascolarctos cinereus*
Konik horse: K*onik polski*
kudzu: *Pueraria lobata*

land crab: *Johngarthia lagostoma*
lemming: *Lemmus lemmus*
lesser black-backed gull: *Larus fuscus*
lesser celandine: *Ranunculus ficaria*
linnet: *Carduelis cannabina*
little egret: *Egretta garzetta*
lyme grass: *Leymus arenarius*

Madagascan periwinkle: *Catharanthus roseus*
marbled white butterfly: *Melanargia galathea*
Maya nut tree: *Brosimum alicastrum*
meadow-grass: *Poa pratensis*
Mediterranean mussel: *Mytilus galloprovincialis*
mesquite, see Mexican thorn

Mexican burweed: *Solanum rostratum*
Mexican thorn: *Prosopis juliflora*
Micronesian megapode: *Megapodius laperouse*
monarch butterfly: *Danaus plexippus*
mongoose: *Herpestes javanicus*
monkey puzzle tree: *Araucaria araucana*
moose: *Alces alces*
Moroccan poppy: *Papaver atlanticum*
mouse-ear hawkweed: *Hieracium pilosella*
muntjac deer: *Muntiacus reevesi*
musk ox: *Ovibos moschatus*
muskrat: *Ondatra zibethica*

New Zealand flatworm: *Arthurdendyus triangulatus*
nightingale: *Luscinia megarhynchos*
Nile monitor: *Varanus niloticus*
Nile perch: *Lates niloticus*
Norfolk pine: *Araucaria heterophylla*
Northern Pacific seastar: *Asterias amurensis*
northern snakehead: *Channa argus*
Norway spruce: *Picea abies*

opossum: *Didelphimorphia*
orang-utan: *Pongo pygmaeus*
ostrich: *Struthio camelus*
otter: *Lutra lutra*
Oxford ragwort: Senecio squalidus

pampas grass: *Cortaderia selloana*
peanut: *Arachis hypogaea*
pectoral sandpiper: *Calidris melanotos*
peregrine falcon: *Falco peregrinus*
pine marten: *Martes martes*
Pinta giant tortoise: *Chelonoidis nigra abingdonii*

Polynesian rat: *Rattus exulans*
prickly pear: *Opuntia monacantha*
purple heron: *Ardea purpurea*
purple loosestrife: *Lythrum salicaria*
purple nut sedge: *Cyperus rotundus*

quoll: *Dasyurus*

raccoon: *Procyon lotor*
red deer: *Cervus elaphus*
red-billed leiothrix: *Leiothrix lutea*
red-necked wallaby: *Macropus rufogriseus*
red squirrel: *Sciurus vulgaris*
reindeer: *Rangifer tarandus*
rhododendron: *Rhododendron ponticum*
ring-necked parakeet: *Psittacula krameri*
robin: *Erithacus rubecula*
roe deer: *Capreolus capreolus*
rook: *Corvus frugilegus*
rosebay willowherb: *Chamerion angustifolium*
royal penguin: *Eudyptes schlegeli*
ruddy duck: *Oxyura jamaicensis*
Russian thistle: *Salsola tragus*

scarce emerald damselfly: *Lestes dryas*
Scottish wildcat: *Felis silvestris grampia*
sea sandwort: *Honckenya peploides*
Seychelles fruit bat: *Pteropus seychellensis*
sheep: *Ovis aries*
shrill carder bee: *Bombus sylvarum*
signal crayfish: *Pacifastacus leniusculus*
smallmouth bass: *Micropterus dolomieu*
snow bunting: *Plectrophenax nivalis*
snowdrop: *Galanthus nivalis*

sockeye salmon: *Oncorhynchus nerka*
song thrush: *Turdus philomelos*
South Georgia pintail: *Anas georgica georgica*
South Georgia pipit: *Anthus antarcticus*
southern giant petrel: *Macronectes giganteus*
southwestern willow flycatcher: *Empidonax traillii extimus*
springtail: *Folsomides parvulus*
sticky groundsel: *Senecio viscosus*
stinging nettle: *Urtica dioica*
stoat: *Mustela erminea*
striped bass: *Morone saxatilis*
Sumatran rhino: *Dicerorhinus sumatrensis*
swamp sago: *Metroxylon sagu*

tamarisk: *Tamarix ramosissima*
thorn apple: *datura stramonium*
Tristan albatross: *Diomedea dabbenena*
trumpet tree: *Cecropia peltata*
tsetse fly: *Glossina palpalis*

vole: *Myodes glareolus*

water buffalo: *Bubalus bubalis*
water hyacinth: *Eichhornia crassipes*
water vole: *Arvicola terrestris*
white-headed duck: *Oxyura leucocephala*
white-tailed eagle: *Haliaeetus albicilla*
white-throated rail: *Dryolimnas cuvieri*
wildebeest: *Connochaetes*
wild boar: *Sus scrofa*
wild rubber: *Hevea brasiliensis*
wolf: *Canis lupus lupus*
wolverine: *Gulo gulo*
wood mouse: *Apodemus sylvaticus*

woolly mammoth: *Mammuthus primigenius*

yellow crazy ant: *Anoplolepis gracilipes*
yellowhammer: *Emberiza citrinella*
yellow trumpetbush: *Tecoma stans*
yellow-wort: *Blackstonia perfoliata*
York groundsel: *Senecio eboracensis*

zebra mussel: *Dreissena polymorpha*

References

Chapter 1: On Green Mountain

1. David Catling and Stedson Stroud, 'The Greening of Green Mountain, Ascension Island', David Catling's Astrobiology, Planetary, and Geobiology Research Page, Department of Earth and Space Sciences, University of Washington, http://faculty.washington.edu/dcatling/Catling2012_GreenMountainSubmitted.pdf
2. Fred Pearce, 'At a Crossroads in the Atlantic', *Geographical*, January 2014, 24.
3. David Wilkinson, 'The Parable of Green Mountain: Ascension Island, Ecosystem Construction and Ecological Fitting', *Journal of Biogeography* 31, no. 1 (22 December 2004): 1–4, doi:10.1046/j.0305-0270.2003.01010.x
4. William Bourne et al., 'A New Sub-fossil Night Heron and a New Genus for the Extinct Rail from Ascension Island', *Ardea* 91 (2003): 45, http://ardea.nou.nu/ardeapdf/a91-045-051.pdf
5. Robin McKie, 'Frigatebird Returns to Nest on Ascension for First Time Since Darwin', *Observer* (UK), 8 December 2012, http://www.theguardian.com/environment/2012/dec/08/frigatebird-returns-to-ascension
6. Roger Huxley, 'Historical Overview of Marine Turtle Exploitation, Ascension Island, South Atlantic', *Marine Turtle Newsletter* 84 (1999): 7–9, http://www.seaturtle.org/mtn/archives/mtn84/mtn84p7.shtml
7. Alan Gray et al., 'The Conservation of the Endemic Vascular Flora of Ascension Island and Threats from Alien Species', *Oryx* 39, no. 4 (October 2005): 1–5, doi:10.1017/S0030605305001092
8. 'Evolution on Islands', Payseur Lab, http://payseur.genetics.wisc.edu/IslandEvolution.htm, accessed January 2014.
9. Ross Wanless et al., 'Predation of Atlantic Petrel Chicks by House Mice on Gough Island', *Animal Conservation* 15, no. 5 (8 May 2012): 472–79, doi:10.1111/j.1469-1795.2012.00534.x
10. John Parkes, *A Feasibility Study for the Eradication of House Mice from Gough Island*, RSPB Research Report No. 34 (Bedfordshire, UK: Royal Society for the Protection of Birds, 2008).

11. Daniel Simberloff, *Invasive Species: What Everyone Needs to Know* (New York: Oxford University Press, 2013).
12. Joseph Mascaro, 'Origins of the Novel Ecosystems Concept,' in *Novel Ecosystems: Intervening in the New Ecological World Order*, ed. Richard Hobbs et al. (Hoboken, NJ: Wiley-Blackwell, 2013), 45–57.
13. Louise Young, *Islands: Portraits of Miniature Worlds* (New York: W.K. Freeman, 1999).
14. Mascaro, 'Origins of the Novel Ecosystems Concept', 45–57.
15. Vanessa Collingridge, *Captain Cook: Obsession and Betrayal in the New World* (London: Ebury, 2002).
16. Christopher Lever, *They Dined on Eland: The Story of the Acclimatization Societies* (London: Quiller Press, 1992).
17. Joseph Mascaro et al., 'Novel Forests Maintain Ecosystem Processes After the Decline of Native Tree Species', *Ecological Monographs* 82 (2012): 221–28.
18. Jeffrey Foster and Scott Robinson, 'Introduced Birds and the Fate of Hawaiian Rainforests', *Conservation Biology* 21, no. 5 (October 2007): 1248–57, doi:10.1111/j.1523-1739.2007.00781.x
19. Andrew P. Dobson, *Conservation and Biodiversity* (New York: Henry Holt, 1998), 195.
20. Peter Vitousek et al., 'Biological Invasion by *Myrica faya* Alters Ecosystem Development in Hawaii', *Science* 238, no. 4828 (6 November 1987), 802, http://www.esf.edu/efb/parry/vitousek.pdf
21. '*Morella faya*', Global Invasive Species Database, http://www.issg.org/database/species/ecology.asp?si=100, accessed February 2014.
22. Dennis O'Dowd et al., 'Invasional "Meltdown" on an Oceanic Island', *Ecology Letters* 6, no. 9 (2003), 812–17, doi:10.1046/j.1461-0248.2003.00512.x
23. 'Yellow Crazy Ants on Christmas Island', Wilderness Society, 23 October 2013, http://www.wilderness.org.au/articles/yellow-crazy-ants-christmas-island
24. Peter Green et al., 'Invasional Meltdown: Invader–Invader Mutualism Facilitates a Secondary Invasion', *Ecology* 92, no. 9 (2011): 1758–68, http://dx.doi.org/10.1890/11-0050.1
25. Veronica Miravete et al., 'How Many and Which Ant Species Are Being Accidentally Moved Around the World?', *Biology Letters* 9, no. 5 (14 August 2013), doi:10.1098/rsbl.2013.0540
26. Meghan Cooling et al., 'The Widespread Collapse of an Invasive

Species: Argentine Ants (*Linepithema humile*) in New Zealand', *Biology Letters* (30 November 2011), doi:10.1098/rsbl.2011.1014
27. Jared Diamond, 'Easter's End', *Discover*, August 1995, 63–9.
28. Terry Hunt, 'Rethinking the Fall of Easter Island,' *American Scientist*, September–October 2006, http://www.americanscientist.org/issues/pub/rethinking-the-fall-of-easter-island/1; Terry Hunt and Carl Lipo, *The Statues That Walked: Unraveling the Mystery of Easter Island* (New York: Free Press, 2011).
29. 'Mike Fay's Pitcairn Journal: Tragic Sighting', *NewsWatch, National Geographic News Blog*, http://newswatch.nationalgeographic.com/2012/04/28/mike-fays-pitcairn-journal-tragic-sighting, accessed November 2013.
30. T.H. Fritts and G.H. Rodda, 'The Brown Tree Snake, an Introduced Pest Species in the Central Pacific Islands', *Endangered Species Technical Bulletin* 14(7): 5–7.
31. Gordon H. Rodda et al., 'The Disappearance of Guam's Wildlife: New Insights for Herpetology, Evolutionary Ecology, and Conservation', *Bioscience* 47, no. 9 (October 1997): 565–74, http://www.jstor.org/discover/10.2307/1313163?uid=2&uid=4&sid=21104657683093
32. '*Boiga irregularis* (brown tree snake)', Global Invasive Species Database, http://www.issg.org/database/species/impact_info.asp?si=54, accessed March 2014.
33. 'Bird Populations in the Northern Mariana Islands Are Being Decimated by Brown Tree Snakes', BirdLife International, presented as part of the BirdLife State of the World's Birds website, http://www.birdlife.org/datazone/sowb/casestudy/400, accessed March 2014.
34. James Legge, '2,000 Dead Mice Dropped on Guam by US Helicopters to Kill Snakes', *Independent* (UK), 3 December 2013, http://www.independent.co.uk/news/world/americas/us-helicopters-drop-painkillerstuffed-dead-mice-over-guam-to-kill-snakes-8980458.html
35. Christoff Kueffer et al., 'A Global Comparison of Plant Invasions on Oceanic Islands', *Perspectives in Plant Ecology, Evolution and Systematics* 12 (2010): 145–61, doi:10.1016/j.ppees.2009.06.002

Chapter 2: New Worlds

1. Tom Griffiths and Libby Robin, *Ecology and Empire: Environmental History of Settler Societies* (Seattle: University of Washington Press, 1997).

2. Peter Savolainen et al., 'A Detailed Picture of the Origin of the Australian Dingo, Obtained from the Study of Mitochondrial DNA', *PNAS (Proceedings of the National Academy of Sciences)* 101, no. 33 (17 August 2004): 12387–90, doi:10.1073/pnas.0401814101
3. David Trigger et al., 'Ecological Restoration, Cultural Preferences and the Negotiation of "Nativeness" in Australia', *Geoforum* 39, no. 3 (May 2008): 1273–83, doi:10.1016/j.geoforum.2007.05.010
4. Adrian Franklin, *Animal Nation: The True Story of Animals and Australia* (Sydney: University of New South Wales Press, 2006), 21.
5. Quoted in William Lines, *Taming the Great South Land: A History of the Conquest of Nature in Australia* (Athens: University of Georgia Press, 1991).
6. Mark Davis, 'Researching Invasive Species 50 Years After Elton: A Cautionary Tale', in *Fifty Years of Invasion Ecology: The Legacy of Charles Elton*, ed. David M. Richardson (Oxford: Blackwell, 2011), 269–76.
7. John McNeill, *Something New Under the Sun: An Environmental History of the Twentieth-Century* (London: Allen Lane, 2000), 254.
8. Alfred W. Crosby, *The Columbian Exchange: Biological and Cultural Consequences of 1492* (Westport, CT: Greenwood Press, 1972).
9. Cecilia Anderung et al., 'Prehistoric Contacts over the Straits of Gibraltar Indicated by Genetic Analysis of Iberian Bronze Age Cattle', *PNAS* 102, no. 24 (14 June 2005): 8431–35, doi:10.1073/pnas.0503396102
10. Christian Kull et al., 'Melting Pots of Biodiversity: Tropical Smallholder Farm Landscapes as Guarantors of Sustainability', *Environment: Science and Policy for Sustainable Development* 55 (March–April 2013): 6–16, http://www.environmentmagazine.org/archives/back%20issues/2013/march-april%202013/melting-pot-full.html
11. Richard Mabey, *Weeds: The Story of Outlaw Plants* (London: Profile Books, 2010), 15.
12. Kathy MacKinnon et al., *The Ecology of Kalimantan: Indonesian Borneo* (Hong Kong: Periplus Editions, 1996).
13. Ivor Noël Hume, 'We Are Starved', *Colonial Williamsburg Journal* (Winter 2007), http://www.history.org/Foundation/journal/Winter07/starving.cfm
14. Paul F. Hendrix et al., 'Pandora's Box Contained Bait: The Global Problem of Introduced Earthworms', *Annual Review of Ecology,*

Evolution, and Systematics 39 (2008): 593–613, doi:10.1146/annurev.ecolsys.39.110707.173426

15. David E. Stannard, *American Holocaust: The Conquest of the New World* (New York: Oxford University Press, 1992), 146.
16. Lucy Burns, 'The Great Cold War Potato Beetle Battle', *BBC News Magazine*, 3 September 2013, http://www.bbc.co.uk/news/magazine-23929124
17. Fred Pearce, *Deep Jungle* (London: Eden Project Books, 2005).
18. Iris Borowy, 'The Other Side of Bio-invasion: The Example of Acclimatization in Germany', in *Invasive and Introduced Plants and Animals: Human Perceptions, Attitudes and Approaches to Management*, ed. Ian Rotherham and Robert Lambert (London: Earthscan, 2011), 153–68.
19. Philip E. Hulme, 'Addressing the Threat to Biodiversity from Botanic Gardens', *Trends in Ecology and Evolution* 26 (2011): 168–74, doi:10.1016/j.tree.2011.01.005
20. 'An Historical Dinner: A Feast of Rarities', *Timaru Herald*, 1910, PapersPast, National Library of New Zealand, http://paperspast.natlib.govt.nz/, accessed March 2014.
21. George M. Thomson, *The Naturalisation of Animals and Plants of New Zealand* (London: Cambridge University Press, 1922), http://archive.org/stream/naturalisationof00thomilia/naturalisation00thomilia_djvu.txt
22. 'An Historical Dinner'.
23. 'American Acclimatization Society', *New York Times*, 15 November 1877, http://query.nytimes.com/mem/archive-free/pdf?_r=1&res=9A03E3D6103FE63BBC4D52DFB767838C669FDE
24. Jacksonville District Aquatic Plant Control Program, US Army Corps of Engineers, http://www.saj.usace.army.mil/Missions/Environmental/InvasiveSpecies/AquaticPlantControlProgram.aspx, accessed March 2014.
25. UNEP Global Environmental Alert Service (GEAS), April 2013, http://www.unep.org/pdf/UNEP_GEAS_APRIL_2013.pdf
26. Fred Pearce, 'All-Out War on the Invader', *New Scientist*, 23 May 1998, 34–8.
27. MacKinnon et al., *The Ecology of Kalimantan*.
28. Martin Dawes, 'Weevil Works Wonders for Lake Victoria', BBC News, 16 December 1999, http://news.bbc.co.uk/2/hi/africa/567884.stm
29. Lawrence M. Kiage and Joyce Obuoyu, 'The Potential Link between

El Nino and Water Hyacinth Blooms in Winam Gulf of Lake Victoria, East Africa: Evidence from Satellite Imagery', *Water Resource Management* 25, no. 14 (November 2011): 3931–45, doi:10.1007/s11269-011-9895-x
30. Zhi Wang et al., 'Large-Scale Utilization of Water Hyacinth for Nutrient Removal in Lake Dianchi in China: The Effects on the Water Quality, Macrozoobenthos and Zooplankton', *Chemosphere* 89 (2012): 1255–61, doi:10.1016/j.chemosphere.2012.08.001
31. 'TED Case Studies: Lake Victoria', Mandala Projects, American University, http://www1.american.edu/ted/victoria.htm, accessed February 2014.
32. Edward O. Wilson, *The Diversity of Life* (London: Allen Lane, 1992), 256.
33. Dirk Verschuren et al., 'History and Timing of Human Impact on Lake Victoria, East Africa', *Proceedings of the Royal Society B* 269 (February 2002): 289–94, doi:10.1098/rspb.2001.1850
34. Frans Witte et al., 'Recovery of Cichlid Species in Lake Victoria: An Examination of Factors Leading to Differential Extinction', *Reviews in Fish Biology and Fisheries* 10, no. 2 (May 2000): 233–41, doi:10.1023/A:1016677515930
35. Paul Robbins, 'Comparing Invasive Networks: Cultural and Political Biographies of Invasive Species', *Geographical Review* 94, no. 2 (2004): 139–56, doi:10.1111/j.1931-0846.2004.tb00164.x
36. 'Mesquite—the Wonder Tree,' *aneyefortexas* (blog), http://aneyefortexas.wordpress.com/2011/03/29/mesquite-the-wonder-tree/, accessed January 2014.
37. 'Success Stories from Sudan,' United Nations Earth Summit+Five, http://www.un.org/esa/earthsummit/sudan.htm, accessed January 2014.
38. Gordon Boy and Arne Witt, *Invasive Alien Plants and Their Management in Africa* (Nairobi: UNEP/GEF, 2013), 43–5.
39. John F. Obiri, 'Invasive Plant Species and Their Disaster-Effects in Dry Tropical Forests and Rangelands of Kenya and Tanzania,' *Journal of Disaster Risk Studies* 3, no. 2 (2011): 417–28, doi:10.4102/jamba.v3i2.39
40. Esther Mwangi and Brent Swallow, '*Prosopis juliflora* Invasion and Rural Livelihoods in the Lake Baringo Area of Kenya', *Conservation and Society* 6, no. 2 (2008): 130–40, doi:10.4103/0972-4923.49207
41. N.M. Pasiecznik et al., 'Putting Knowledge on *Prosopis* into Use

in Kenya', Pioneering Advances in 2006, KEFRI (Nairobi, Kenya) and HDRA (Coventry, UK), http://www.jatropha.pro/PDF%20 bestanden/ProsopisKenyaSummaryReport.pdf

Chapter 3: All at Sea

1. Fred Pearce, 'How the Soviet Seas Were Lost', *New Scientist*, 11 November 1995, 38–42.
2. Yu Zaitsev and B.G. Alexandrov, 'Recent Man-Made Changes in the Black Sea Ecosystem', in *Sensitivity to Change: Black Sea, Baltic Sea and North Sea*, NATO ASI Series, ed. Emin Özsoy and Alexander Mikaelyan (Dordrecht, Netherlands: Kluwer Academic, 1997), 25–31.
3. Bella Galil et al., 'First Record of *Mnemiopsis leidyi* A. Agassiz, 1865 (Ctenophora; Lobata; Mnemiidae) off the Mediterranean Coast of Israel', *Aquatic Invasions* 4, no. 2 (January 2009): 256–62, doi:10.3391/ai.2009.4.2.8
4. Zaitsev and Alexandrov, 'Recent Man-Made Changes in the Black Sea Ecosystem'.
5. 'Ballast Water—A Pathway for Aquatic Invasive Species', NOAA Fisheries Service, http://www.habitat.noaa.gov/pdf/best_management_practices/fact_sheets/Ballast%20Water%20Factsheet.pdf, accessed October 2013.
6. Hanno Seebens et al., 'The Risk of Marine Bioinvasion Caused by Global Shipping', *Ecology Letters* 16, no. 6 (June 2013): 782–90, doi:10.1111/ele.12111
7. 'The IMO Ballast Water Management Convention', GloBallast Partnerships, http://globallast.imo.org/index.asp?page=mepc.htm, accessed January 2014.
8. Fred Dobbs and Andrew Rogerson, 'Ridding Ships' Ballast Water of Micro-organisms', *Environmental Science & Technology*, 15 June 2005, 259A–264A.
9. Fred Pearce, 'Dead in the Water', *New Scientist*, 4 February 1995, 26–31.
10. Susan Williams and Edwin Grosholz, 'The Invasive Species Challenge in Estuarine and Coastal Environments: Marrying Management and Science', *Estuaries and Coasts* 31, no. 1 (February 2008): 3–20, doi:10.1007/s12237-007-9031-6
11. Sales copy for Alexandre Meinesz, *Killer Algae: The True Tale of a Biological Invasion* (Chicago: University of Chicago Press, 1999), http://www.press.uchicago.edu/ucp/books/book/chicago/K/bo3644960.html

12. Jean Jaubert et al., 'Re-evaluation of the Extent of *Caulerpa taxifolia* Development in the Northern Mediterranean Using Airborne Spectrographic Sensing', *Marine Ecology Progress Series* 263 (2003): 75–82.
13. '*Caulerpa taxifolia* (Alga)', Global Invasive Species Database, http://www.issg.org/database/species/ecology.asp?si=115&fr=1&sts=tss&lang=EN, accessed January 2014.
14. David K.A. Barnes, 'Biodiversity: Invasions by Marine Life on Plastic Debris', *Nature* 416 (25 April 2002): 808–9, doi:10.1038/416808a
15. David K.A. Barnes, 'Natural and Plastic Flotsam Stranding in the Indian Ocean', in *The Effects of Human Transport on Ecosystems: Cars and Planes, Boats and Trains*, ed. John Davenport and Julia L. Davenport (Dublin: Royal Irish Academy, 2004), 193–205, http://5gyres.org/media/Flotsam_in_the_Indian_Ocean.pdf
16. Jeff Barnard, 'Invasive Species Ride Tsunami Debris to US Shore', *Christian Science Monitor*, 9 June 2012, http://www.csmonitor.com/USA/Latest-News-Wires/2012/0609/Invasive-species-ride-tsunami-debris-to-US-shore
17. Fred Pearce, 'Zoologger: The Mysterious Crabs of Ascension Island', *New Scientist Online*, 27 September 2013, http://www.newscientist.com/article/dn24287
18. Ruprecht Jaenicke, 'Abundance of Cellular Material and Proteins in the Atmosphere', *Science* 308, no. 5718 (1 April 2005): 73, doi:10.1126/science.1106335
19. S.M. Burrows et al., 'Bacteria in the Global Atmosphere – Part 1: Review and Synthesis of Literature Data for Different Ecosystems', *Atmospheric Chemistry and Physics* 9 (January 2009): 10777–827, doi:10.5194/acp-9-9263-2009
20. Fred Pearce, 'The Long Strange Journey of Earth's Traveling Microbes', *Yale Environment 360*, 22 August 2011, http://e360.yale.edu/feature/the_long_strange_journey_of_earths_traveling_microbes/2436
21. Robin McKie, 'Deadly Dust "Brought Foot and Mouth" Here', *Observer* (UK), 8 September 2001, http://www.theguardian.com/uk/2001/sep/09/footandmouth.robinmckie
22. 'Diversity in the Air', *Astrobiology*, 27 December 2006, http://www.astrobio.net/pressrelease/2188/diversity-in-the-air
23. Lily Young and Jerry Kukor, 'Bioaerosols', lecture notes, Environmental and Pollution Microbiology course, Rutgers University

Department of Environmental Sciences, http://envsci.rutgers.edu/510/pdf/Bioaerosols_2010.pdf, accessed January 2014.
24. Charles Elton, *The Ecology of Invasions by Animals and Plants* (1958; reprinted, Chicago: University of Chicago Press, 2000), 44.

Chapter 4: Welcome to America

1. Peter Barnes, 'Japan's Botanical Sunrise: Plant Exploration Around the Meiji Restoration', *Curtis's Botanical Magazine* 18, no. 2 (May 2001): 117–31, doi:10.1111/1467-8748.00300
2. Derek H. Alderman, 'Channing Cope and the Making of the Miracle Vine', *Geographical Review* 94, no. 2 (April 2004): 157–77, http://www.jstor.org/stable/30033969
3. Marilyn Witkamp et al., 'Accumulation and Biota in a Pioneering Ecosystem of Kudzu Vine at Copperhill, Tennessee', *Journal of Applied Ecology* 3 (1966): 383–91.
4. 'Fact Sheet: Kudzu', Plant Conservation Alliance's Alien Plant Working Group, http://www.nps.gov/plants/alien/fact/pdf/pumo1.pdf, accessed January 2014.
5. Irwin N. Forseth and Anne F. Innis, 'Kudzu (*Pueraria montana*): History, Physiology, and Ecology Combine to Make a Major Ecosystem Threat', *Critical Reviews in Plant Sciences* 23, no. 5 (January 2004): 401–13, doi:10.1080/07352680490505150
6. Anna E. Eskridge and Derek H. Alderman, 'Alien Invaders, Plant Thugs, and the Southern Curse: Framing Kudzu as Environmental Other Through Discourses of Fear', *Southeastern Geographer* 50, no. 1 (Spring 2010): 110–29, doi:10.1353/sgo.0.0073
7. Various articles in the *New York Times*, http://www.nytimes.com, accessed November 2013.
8. Ashley Poplin and Amanda Hodges, 'Bean Plataspid: *Megacopta cribraria* (Fabricius) (Insecta: Hemiptera: Heteroptera: Plataspidae)', University of Florida IFAS Extension, http://edis.ifas.ufl.edu/in939, accessed November 2013.
9. David Fairchild, *Garden Islands of the Great East: Collecting Seeds from the Philippines and Netherlands India in the Junk 'Chêng ho'* (New York: C. Scribner's Sons, 1943).
10. Mark Alfred Carleton, 'Adaptation of the Tamarisk for Dry Lands', *Science* 39 (1914): 692–94, https://archive.org/details/jstor-1639861
11. Matthew Chew, 'The Monstering of Tamarisk: How Scientists Made

a Plant into a Problem', *Journal of the History of Biology* 42 (2009): 231–66, doi:10.1007/s10739-009-9181-4

12. J.S. Gatewood et al., *Use of Water by Bottom-Land Vegetation in Lower Safford Valley, Arizona*, Geological Survey Water-Supply Paper 1103 (Washington, DC: US Geological Survey, 1950), http://pubs.usgs.gov/wsp/1103/report.pdf
13. Todd Hartman, 'With Luck, Beetles May Spill Salt Cedars', *Rocky Mountain News* (Denver, CO), 18 August 2001, http://www.highbeam.com/doc/1G1–77344258.html
14. 'Salt Cedar', Plant Conservation Alliance's Alien Plant Working Group, http://www.nps.gov/plants/ALIEN/fact/tama1.htm, accessed October 2013.
15. Charles R. Hart et al., 'Saltcedar Control and Water Salvage on the Pecos River, Texas, 1999–2003', *Journal of Environmental Management* 75 (2005): 399–409, doi:10.1614/IPSM-09-009.1
16. Martin Schlaepfer et al., 'The Potential Conservation Value of Non-native Species', *Conservation Biology* 25, no. 3 (June 2011): 428–37, doi:10.1111/j.1523-1739.2010.01646.x
17. Mark K. Sogge et al., 'Saltcedar and Southwestern Willow Flycatchers: Lessons from Long-Term Studies in Central Arizona', US Geological Survey, http://sbsc.wr.usgs.gov/cprs/research/projects/swwf/Reports/Sogge_et_al_Saltcedar_and_SWWF_proceedings_with_cover.pdf
18. Gordon Whitney, *From Coastal Wilderness to Fruited Plain: A History of Environmental Change in Temperate North America from 1500 to the Present* (Cambridge, UK: Cambridge University Press, 1994), 289–90.
19. Daniel Q. Thompson et al., 'Spread, Impact, and Control of Purple Loosestrife (*Lythrum salicaria*) in North American Wetlands', US Geological Survey, http://www.npwrc.usgs.gov/resource/plants/loosstrf, accessed November 2013.
20. Michael A. Treberg and Brian C. Husband, 'Relationship Between the Abundance of *Lythrum salicaria* (Purple Loosestrife) and Plant Species Richness Along the Bar River, Canada', *Wetlands* 19, no. 1 (1999): 118–25, doi:10.1007/BF03161740
21. Robert Devine, 'The Cheatgrass Problem', *Atlantic*, May 1993, 40–50.
22. John Maerz et al., 'Declines in Woodland Salamander Abundance Associated with Non-native Earthworm and

Plant Invasions', *Conservation Biology* 23 (2009): 975–81, doi:10.1111/j.1523-1739.2009.01167.x

23. Anne Minard, 'Researchers Build a Case for Earthworm's Slimy Reputation', *New York Times*, 28 October 2003, http://www.nytimes.com/2003/10/28/science/researchers-build-a-case-for-earthworm-s-slimy-reputation.html
24. Cindy Hale, 'Evidence for Human-Mediated Dispersal of Exotic Earthworms: Support for Exploring Strategies to Limit Further Spread', *Molecular Ecology* 17 (21 February 2008): 1165–67, doi:10.1111/j.1365-294X.2007.03678.x
25. 'Making a Hash of the Ash', *Economist*, 16 March 2013, http://www.economist.com/news/united-states/21573591-flying-scourge-widens-its-reach-making-hash-ash
26. 'Zebra Mussel', Great Lakes Science Center, US Geological Survey, http://pbadupws.nrc.gov/docs/ML1126/ML112640036.pdf, accessed January 2014.
27. Diana Hamilton et al., 'Predation of Zebra Mussels by Diving Ducks: An Exclosure Study', *Ecology* 75 (January 1994): 521–31, doi:10.2307/1939555
28. Daniel Simberloff, foreword to *The Ecology of Invasions by Animals and Plants*, by Charles Elton (Chicago: University of Chicago Press, 2000).
29. T.J. Pignataro, 'Sturgeon Battles Back to Repopulate Lake Erie, Lower Niagara River', *Buffalo (NY) News*, 7 September 2013, http://www.buffalonews.com/city-region/sturgeon-battles-back-to-repopulate-lake-erie-lower-niagara-river-20130907
30. Nancy Connelly, 'Economic Impacts of Zebra Mussels on Drinking Water Treatment and Electric Power Generation Facilities', *Environmental Management* 40, no. 1 (August 2007): 105–12, doi:10.1007/s00267-006-0296-5
31. 'Zebra Mussel'.
32. Dan Egan, 'Asian Carp DNA Found in Wisconsin's Lake Michigan Waters for First Time', *Journal Sentinel* (Milwaukee), 5 November 2013, http://www.jsonline.com/news/wisconsin/asian-carp-dna-found-in-sturgeon-bay-b99135592z1-230687101.html
33. Michael Richardson, 'Hitchhiking Species a Threat Worldwide', *International Herald Tribune*, 28 August 2002.
34. Sherri Graves and Arthur Shapiro, 'Exotics as Host Plants of the California Butterfly Fauna', *Biological Conservation* 110 (April 2003): 413–33, doi:10.1016/S0006-3207(02)00233-1

35. Andrew Cohen and James Carlton, 'Accelerating Invasion Rate in a Highly Invaded Estuary', *Science* 279, no. 5350 (23 January 1998): 555.
36. James Carlton et al., 'Remarkable Invasion of San Francisco Bay (California, USA) by the Asian Clam *Potamocorbula amurensis*: 1. Introduction and Dispersal', *Marine Ecology Progress Series* 66 (1990): 81–94, doi:10.3354/meps066081
37. 'Corbula amurensis', *The Exotics Guide: Non-Native Marine Species of the North American Pacific Coast*, http://www.exoticsguide.org/corbula_amurensis, accessed October 2013.
38. Doug Cordell, 'San Francisco Bay NWRC: International "Ramsar" Recognition of San Francisco Bay Wetlands Celebrated', US Fish and Wildlife Service Field Notes, 13 May 2013, http://www.fws.gov/fieldnotes/regmap.cfm?arskey=33792, accessed October 2013.
39. 'Thousands of Pythons Invading the Florida Everglades', *America Now*, 2011, http://www.americanownews.com/story/15870487/thousands-of-giant-pythons-invading-the-us
40. Michael E. Dorcas et al., 'Severe Mammal Declines Coincide with Proliferation of Invasive Burmese Pythons in Everglades National Park', *PNAS* 109, no. 7 (14 February 2012): 2418–22, doi:10.1073/pnas.1115226109
41. Carla J. Dove et al., 'Consumption of Bird Eggs by Invasive Burmese Pythons in Florida', *IRCF Reptiles & Amphibians: Conservation and Natural History* 19, no. 1 (March 2012): 64–66, http://www.ircf.org/journal/issue-191-mar-2012/
42. 'The Everglades Historical Timeline', *Water's Journey: Everglades*, http://www.theevergladesstory.org/history/history.php, accessed October 2013.
43. Theodore (Ted) Mosquin, 'Management Guidelines for Invasive Alien Species in Canada's National Parks', Ecospherics International, http://www.ecospherics.net/AlienSpecnew.htm, accessed August 2014.
44. 'Invasive Species Definition Clarification and Guidance White Paper', Definitions Subcommittee of the Invasive Species Advisory Committee, 'approved by ISAC April 27, 2006', National Invasive Species Council, http://www.seagrant.sunysb.edu/ais/pdfs/definititions.pdf
45. Osvaldo E. Sala et al., 'Global Biodiversity Scenarios for the Year 2100', *Science* 287 (2000): 1770–74, doi:10.1126/science.287.5459.1770

Chapter 5: Britain: A Nation Tied in Knotweed

1. Ann Townsend, 'Japanese Knotweed: A Reputation Lost', *Arnoldia* (Fall 1997): 13–19, http://arnoldia.arboretum.harvard.edu/pdf/articles/471.pdf
2. As quoted in 'The Making of Monsters', *Kew*, magazine of the Royal Botanic Gardens, Kew, London, Spring 2001, 24–6.
3. Richard Mabey, *Flora Britannica* (London: Chatto & Windus, 1996).
4. 'BBC One Show – Japanese Knotweed Property Damage', YouTube, http://www.youtube.com/watch?v=Yu-Y68W4K5Y, accessed November 2013.
5. Lisa Campbell, 'Japanese Knotweed: The Scourge That Could Sink Your House Sale', *Observer* (UK), 8 September 2012, http://www.theguardian.com/money/2012/sep/08/japanese-knotweed-house-sale
6. 'Invasive Non-native Species', *Postnote*, no. 303 (April 2008), http://www.parliament.uk/documents/post/postpn303.pdf
7. F. Williams et al., *The Economic Cost of Invasive Non-Native Species on Great Britain* (Wallingford, UK: CAB International, November 2010), www.cabi.org
8. David Derbyshire, 'The Dreaded Alien Eating Your Garden and Home … but Don't Dare Try to Kill the Japanese Knotweed', *Daily Mail* (UK), 9 July 2013, http://www.dailymail.co.uk/news/article-2358599/Alien-eating-garden-home--And-dont-dare-try-kill-dreaded-Japanese-knotweed.html
9. John Bailey, 'The Rise and Fall of Japanese Knotweed', in *Invasive and Introduced Plants and Animals: Human Perceptions, Attitudes and Approaches to Management*, ed. Ian Rotherham and Robert Lambert (London: Earthscan, 2011), 221–32.
10. Mabey, *Flora Britannica*.
11. Ian Rotherham, 'History and Perception in Animal and Plant Invasions – the Case of Acclimatization and Wild Gardeners', in *Invasive and Introduced Plants and Animals*, 233–48.
12. Richard Mabey, *Weeds: The Story of Outlaw Plants* (London: Profile Books, 2010), 231.
13. Richard Drayton, *Nature's Government: Science, Imperial Britain, and the 'Improvement' of the World* (New Haven, CT: Yale University Press, 2000), 191; 'Rhododendron Dell', Kew Royal Botanic Gardens, http://www.kew.org/visit-kew-gardens/garden-attractions-A-Z/rhododendron-dell.htm, accessed December 2013.

14. 'Rhododendron', Eden Project, http://www.edenproject.com/shop/Rhododendron-Rhododendron-Trewithen-Hybrid.aspx, accessed December 2013.
15. Rotherham, 'History and Perception in Animal and Plant Invasions', 233–48.
16. Aurelio Mayo et al., 'Positive Effects of an Invasive Shrub on Aggregation and Abundance of a Native Small Rodent', *Behavioural Ecology* 24 (2013): 759–67.
17. 'Himalayan Balsam', statement from the BBKA (British Beekeepers' Association), July 2011, http://www.bbka.org.uk/files/pressreleases/19-07-2011_bbka_statement_on_himalayan_balsam_1311085195_1374670518.pdf, accessed December 2013.
18. 'Online Atlas of the British & Irish Flora', Biological Records Centre, http://www.brc.ac.uk/plantatlas/, accessed August 2014.
19. Rob Marrs et al., 'Native Dominants in British Woodland – a Potential Cause of Reduced Species-Richness?', *New Journal of Botany* 3, no. 3 (2013): 156–68, doi:10.1179/2042349713Y.0000000028
20. David Pearman and Alex Lockton, 'Alien Invaders?', Botanical Society of Britain and Ireland, January 2007, http://www.bsbi.org.uk/alien_invaders_.html
21. Michael Pocock and Darren Evans, '"Citizen Scientists" Help Track Alien Invader of Conker Trees', University of Hull, School of Biological, Biomedical and Environmental Sciences, http://www2.hull.ac.uk/science/bbes/news/citizenstrackconkertreeinvader.aspx, accessed January 2014.
22. Helen Briggs, 'The Area of Outstanding Beauty Remade by Man',. BBC News, 25 October 2013, http://www.bbc.co.uk/news/science-environment-24239440
23. Andrew Leslie, 'The Ecology and Biodiversity Value of Sycamore (*Acer pseudoplatanus*) with Particular Reference to Great Britain', *Scottish Forestry* 59, no. 3 (2005): 19–26.
24. 'Position Statement on Buddleia and Its Planting in the UK', Butterfly Conservation, October 2012, http://butterfly-conservation.org/files/bc_position_statement_buddleia_nov2012.pdf
25. Stephen Trudgill, 'A Requiem for the British Flora? Emotional Biogeographies and Environmental Change', *Area* 40, no. 1 (March 2008): 99–107, doi:10.1111/j.1475-4762.2008.00790.x
26. Mabey, *Weeds.*

27. Peter Coates, 'Over Here: American Animals in Britain', in *Invasive and Introduced Plants and Animals*, 39–54.
28. 'Top Ten Most Wanted Foreign Species', Environment Agency (UK) news release, 3 August 2006, http://webarchive.nationalarchives.gov.uk
29. George Monbiot, 'How British Nationalists Got Their Claws into My Crayfish', *Guardian Online* (UK), 2 October 2009, http://www.theguardian.com/environment/georgemonbiot/2009/oct/01/crayfish-bnp; 'George Monbiot – Death to the Usurper', *21st Century British Nationalism*, http://leejohnbarnes.blogspot.co.uk/2009/09/george-monbiot-death-to-usurper.html, accessed October 2013.
30. 'Mitten Crabs in the Thames – Gordon Ramsay,' YouTube, http://www.youtube.com/watch?v=5yp9rZBhrVs, accessed October 2013.
31. Lucy Siegle, 'Invasive Non-native Species: Attack of the Aliens', *Observer* (UK), 15 January 2012.
32. 'Edible Dormice', Natural England, http://www.naturalengland.org.uk/ourwork/regulation/wildlife/species/edibledormice.aspx, accessed October 2013.
33. Richard Creasey, 'Invasion of the Glis Glis', *Mail Online*, 23 September 2006, http://www.dailymail.co.uk/news/article-406658/Invasion-glis-glis.html
34. 'Managing Wildlife', *Nature Features*, BBC, 24 May 2013, http://www.bbc.co.uk/nature/22173289; Alexander Lees and Diana Bell, 'A Conservation Paradox for the 21st Century: The European Wild Rabbit *Oryctolagus cuniculus*, an Invasive Alien and an Endangered Native Species', *Mammal Review* 38 (October 2008): 304–20, doi:10.1111/j.1365-2907.2008.00116.x
35. Oliver Rackham, *The History of the Countryside* (London: J.M. Dent, 1986), 96.
36. Christopher Lever, *The Naturalized Animals of Britain and Ireland* (London: New Holland, 2009).
37. Ibid.
38. Steven Morris, 'Animal activists vow to stop planned wild boar cull in Forest of Dean', *Guardian*, 21 August 2014, http://www.theguardian.com/environment/2014/aug/21/wild-boar-forest-of-dean-activists-vow-to-stop-cull
39. George Monbiot, *Feral: Searching for Enchantment on the Frontiers of Rewilding* (London: Allen Lane, 2013), 95.
40. Christopher Lever, *The Naturalized Animals of Britain and Ireland.*

41. Michael McCarthy, 'The decline and fall of the Peak District wallabies', *Independent* (UK), 20 February 2013.
42. Steven McKenzie, '"Send me tails of red squirrels"', BBC News Online, 29 April 2009, http://news.bbc.co.uk/1/hi/scotland/highlands_and_islands/8023283.stm
43. 'Country Diary', *Manchester Guardian*, 11 December 1912, http://www.theguardian.com/environment/2012/dec/09/grey-squirrel-welcome-1912-archive
44. 'Otter Comeback is Good News for Water Voles, Bad News for Mink', *Wildlife Extra*, http://www.wildlifeextra.com/go/news/news-otterincrease.html#cr, accessed October 2013.
45. Christopher Lever, *The Naturalized Animals of Britain and Ireland.*
46. F. Williams et al., *The Economic Cost of Invasive Non-Native Species on Great Britain.*

Chapter 6: Ecological cleansing

1. Rory Carroll, 'Township poor of South Africa risk life and limb in fight against plant invaders', *The Guardian*, 1 November 2003, http://www.theguardian.com/world/2003/nov/01/southafrica.rorycarroll
2. 'Working for Water: A South African Sustainability Case', United Nations Environment Programme, http://www.unep.org/training/programmes/Instructor%20Version/Part_3/readings/WfW_case.pdf, accessed September 2013
3. Sue Armstrong, 'Rare plants protect Cape's water supplies', *New Scientist* 145, no. 1964 (11 February 1995), 8.
4. Michelle Aitken et al., 'Living with Alien Invasives: The Political Ecology of Wattle in the Eastern Highveld Mpumalanga, South Africa', *Études Océan Indien*, 42–43 (2009), 115–41, http://www.academia.edu/4737610/Aitken
5. Megan Nowell, 'Determining the hydrological benefits of clearing invasive alien vegetation on the Agulhas Plain, South Africa', master's thesis, University of Stellenbosch, 2011, http://hdl.handle.net/10019.1/6855
6. Christian Kull et al, 2011, 'Adoption, Use and Perception of Australian Acacias Around the World', *Diversity and Distributions*, 17 (2011), 822–36, doi:10.1111/j.1472-4642.2011.00783.x
7. Brian van Wilgen et al., 2012, 'An Assessment of the Effectiveness of a Large, National-Scale Invasive Alien Plant Control Strategy in

South Africa', *Biological Conservation*, 148, no. 1 (2012), 28–38, doi:10.1016/j.biocon.2011.12.035

8. 'South Georgia Completes First Phase of Reindeer Eradication: 1,900 Animals', *Merco Press* (Montevideo), 19 March 2013, http://en.mercopress.com/2013/03/19/south-georgia-completes-first-phase-of-reindeer-eradication-1.900-animals'; 'South Georgia prepares to cull its invasive reindeer', BBC News Online, 9 January 2013, http://www.bbc.co.uk/news/world-20962250
9. 'Environment Documents', South Georgia and South Sandwich Islands, http://www.sgisland.gs/index.php/(g)Environment_Documents, accessed March 2014.
10. 'Rodent Eradication', UK Overseas Territories Conservation Forum, http://www.ukotcf.org/infoDB/infoSourcesDetail2.cfm?refID=307, accessed March 2014.
11. William Sobey, 'Macquarie Island: The Introduction of the European Rabbit Flea *Spilopsyllus cuniculi (Dale)* as a Possible Vector for Myxomatosis', *Journal of Hygiene* 71, no. 2 (1973): 299–308, http://www.ncbi.nlm.nih.gov/pmc/articles/PMC2130493/
12. 'Lessons Learned from Devastating Effects of Cat Eradication on Macquarie Island', 13 January 2009, Australian Antarctic Division, http://www.antarctica.gov.au/news/2009/lessons-learned-from-devastating-effects-of-cat-eradication-on-macquarie-island
13. Dana M. Bergstrom et al., 'Indirect Effects of Invasive Species Removal Devastate World Heritage Island', *Journal of Applied Ecology* 46 (2009): 73–81, doi: 10.1111/j.1365-2664.2008.01601.x
14. *Macquarie Island Pest Eradication Plan, Part A: Overview*, Australian Government, Department of the Environment, Water, Heritage and the Arts, 2007, http://www.parks.tas.gov.au/file.aspx?id=6743
15. 'Poison Baits Kill Over 400 Macquarie Island Birds', *ABC Hobart*, 22 October 2010, http://www.abc.net.au/news/stories/2010/10/22/3046031.htm?site=hobart
16. John Reid, 'Wrecking Macquarie Island to Save It', *Quadrant Online*, 11 September 2012, http://quadrant.org.au/opinion/doomed-planet/2012/09/wrecking-macquarie-island-to-save-it
17. Piero Genovesi and Laura Carnevali, 'Invasive Alien Species on European Islands: Eradications and Priorities for Future Work', in *Island Invasives: Eradication and Management*, ed. C.R. Veitch et al. (Gland, Switzerland: International Union for Conservation of Nature and Natural Resources, 2011), 5–8; Daniel Simberloff, 'Biological

Invasions: Prospects for Slowing a Major Global Change', *Elemanta* 1 (2013), doi:10.12952/journal.elementa.000008

18. 'American Muskrats Plague Bohemia', *New York Times*, 19 September 1915, http://query.nytimes.com/mem/archive-free/pdf?res=F10E1EFF3F5512738FDDA00994D1405B858DF1D3
19. Charles Elton, *The Ecology of Invasions by Animals and Plants* (1958; reprinted, Chicago: University of Chicago Press, 2000).
20. 'Mink and Water Vole', GB Non-native Species Secretariat, http://www.nonnativespecies.org/index.cfm?pageid=149, accessed January 2014.
21. C.C. Kessler, 'Eradication of Feral Goats and Pigs and Consequences for Other Biota on Sarigan Island, Commonwealth of the Northern Mariana Islands', in *Turning the Tide: The Eradication of Invasive Species*, ed. C.R. Veitch and M.N. Clout, Occasional Paper of the IUCN Species Survival Commission no. 27 (Gland, Switzerland: IUCN, 2002), https://portals.iucn.org/library/efiles/documents/ssc-op-028.pdf
22. Ibid.
23. *Greening the Bay: Financing Wetland Restoration in San Francisco Bay* (Oakland, CA: Save the Bay, 2007), http://www.savesfbay.org/greeningthebay, accessed January 2014.
24. Susan Williams and Edwin Grosholz, 'The Invasive Species Challenge in Estuarine and Coastal Environments: Marrying Management and Science', *Estuaries and Coasts* 31, no. 1 (February 2008): 3–20, doi:10.1007/s12237-007-9031-6
25. Jen McBroom, *Clapper Rail Surveys for the San Francisco Estuary Invasive Spartina Project* (Oakland, CA: Coastal Conservancy, 2012), http://www.spartina.org/project_documents/revegetation_program/CLRA%20Report%202012.pdf
26. 'Clapper Rail', in *The State of Birds San Francisco Bay 2011*, PRBO Conservation Science and the San Francisco Bay Joint Venture, http://data.prbo.org/sfstateofthebirds/index.php?page=clapper-rail, accessed October 2013.
27. 'Prickly Pear', Department of Agriculture, Fisheries and Forestry, Biosecurity Queensland, 2013, http://www.daff.qld.gov.au/__data/assets/pdf_file/0007/76606/IPA-Prickly-Pear-Control-PP29.pdf, accessed October 2013.
28. David Fickling, 'Australia's Plague Ready to Leap Again', *Guardian* (UK), 22 November 2002, http://www.theguardian.com/world/2002/nov/22/australia.davidfickling

29. Richard Shine and J. Sean Doody, 'Invasive Species Control: Understanding Conflicts Between Researchers and the General Community', *Frontiers in Ecology and the Environment* 9, no. 7 (September 2011): 400–06, doi:10.1890/100090
30. Richard Shine, 'The Ecological Impact of Invasive Cane Toads (*Bufo marinus*) in Australia', *Quarterly Review of Biology* 85, no. 3 (September 2010): 253–91, doi:10.1086/655116
31. John Ewel, 'Case Study: Hole-in-the-Donut, Everglades', in *Novel Ecosystems: Intervening in the New Ecological World Order*, ed. Richard Hobbs et al. (Hoboken, NJ: Wiley-Blackwell, 2013), 11–15.
32. Violeta Muñoz-Fuentes et al., 'Hybridization Between White-Headed Ducks and Introduced Ruddy Ducks in Spain', *Molecular Ecology* 16 (2007): 629–38, doi:10.1111/j.1365-294X.2006.03170.x
33. 'Ruddy Ducks Love Rare Cousins to Death', *New Scientist*, 6 February 1993, 10.
34. Jennifer Kross, 'Waterfowl Hybrids', Ducks Unlimited, 2006, http://www.ducks.org/conservation/waterfowl-biology/waterfowl-hybrids

Chapter 7: Myths of the Aliens

1. Charles Elton, *Voles, Mice and Lemmings: Problems in Population Dynamics* (Oxford, UK: Oxford University Press, 1942).
2. Richard Southwood and J.R. Clarke, 'Charles Sutherland Elton', *Biographical Memoirs of Fellows of the Royal Society* 45 (November 1, 1999): 129–46, http://rsbm.royalsocietypublishing.org/content/45/129.abstract
3. Charles Elton, *The Ecology of Invasions by Animals and Plants* (1958; reprinted, Chicago: University of Chicago Press, 2000), 15.
4. Michael Barbour, 'California Landscapes Before the Invaders', California Exotic Pest Plant Council, 1996 symposium proceedings, http://www.cal-ipc.org/symposia/archive/pdf/1996_symposium_proceedings1837.pdf
5. Stephen Jay Gould, 'An Evolutionary Perspective on Strengths, Fallacies and Confusions in the Concept of Native Plants', *Arnoldia* (Spring 1998): 3–10, http://arnoldia.arboretum.harvard.edu/pdf/articles/483.pdf
6. Bob Holmes, 'Day of the Sparrow', *New Scientist*, 27 June 1998, 32–5.
7. Carolyn King, *Immigrant Killers: Introduced Predators and the Conservation of Birds in New Zealand* (Oxford, UK: Oxford University Press, 1984).

8. Don Schmitz and Daniel Simberloff, 'Biological Invasions: A Growing Threat', *Issues in Science and Technology* 13, no. 4 (Summer 1997): 33–40, http://issues.org/13-4/schmit/
9. Daniel Simberloff, 'The Rise of Modern Invasion Biology and American Attitudes Towards Introduced Species', in *Invasive and Introduced Plants and Animals: Human Perceptions, Attitudes and Approaches to Management*, ed. Ian Rotherham and Robert Lambert (London: Earthscan, 2011), 121–36.
10. Petr Pysek et al., 'Geographical and Taxonomic Biases in Invasion Ecology,' *Trends in Ecology and Evolution* 23, no. 5 (2008): 237–44, doi:10.1016/j.tree.2008.02.002
11. Daniel Simberloff, *Invasive Species: What Everyone Needs to Know* (New York: Oxford University Press, 2013).
12. J.L. Ruesink et al., 'Changes in Productivity Associated with Four Introduced Species: Ecosystem Transformation of a "Pristine" Estuary', *Marine Ecology Progress Series* 311 (2006): 203–15, http://www.int-res.com/articles/theme/m311_TS.pdf
13. LeRoy Holm, *The World's Worst Weeds: Distribution and Biology* (Honolulu: University Press of Hawaii, 1977).
14. 'Statistics and Facts', GB Non-Native Species Secretariat, http://www.nonnativespecies.org/index.cfm?pageid=258, accessed January 2014.
15. David Wilcove et al., 1998, 'Quantifying Threats to Imperiled Species in the United States', *BioScience* 48 (1998): 607–15.
16. Mark Davis, 'Researching Invasive Species 50 Years After Elton: A Cautionary Tale', in *Fifty Years of Invasion Ecology*, ed. David M. Richardson (Oxford, UK: Blackwell, 2011), 269–74.
17. Ibid.
18. 'Statistics and Facts', GB Non-Native Species Secretariat.
19. Miguel Clavero and Emili García-Berthou, 'Invasive Species Are a Leading Cause of Animal Extinctions', *Trends in Ecology and Evolution* 20, no. 3 (March 2005): 110, doi.org/10.1016/j.tree.2005.01.003
20. Jessica Gurevitch and Dianna K. Padilla, 'Are Invasive Species a Major Cause of Extinctions?', *Trends in Ecology and Evolution* 19, no. 9 (September 2004): 470–74, doi:10.1016/j.tree.2004.07.005
21. R.K. Didham et al., 'Are Invasive Species the Drivers of Ecological Change?', *Trends in Ecology and Evolution* 20, no. 9 (September 2005): 470–74, doi:10.1016/j.tree.2005.07.006
22. David Pimentel et al., 'Economic and Environmental Threats of

Alien Plant, Animal, and Microbe Invasions', *Agriculture Ecosystems and Environment* 84, no. 1 (January 2001): 1–20, doi:10.1016/S0167-8809(00)00178-X

23. David Pimentel et al., 'Update on the Environmental and Economic Costs Associated with Alien-Invasive Species in the United States', *Ecological Economics* 52 (January 2005): 273–88, doi:10.1016/j.ecolecon.2004.10.002
24. Sue Armstrong, 'Rare Plants Protect Cape's Water Supplies', *New Scientist* 145, no. 1964, 11 February 1995, 8.
25. Working for Water: A South African Sustainability Case', United Nations Environment Programme, http://www.unep.org/training/programmes/Instructor%20Version/Part_3/readings/WfW_case.pdf, accessed September 2013; 'Cape Town's Biodiversity Is Under Threat', Invasive Species South Africa, http://www.invasives.org.za/v2/index.php?option=com_k2&view=item&id=126:cape-town%25E2%2580%2599s-biodiversity-is-under-threat&Itemid=60, accessed January 2014.
26. Dov F. Sax and Steven D. Gaines, 'Species Invasions and Extinction: The Future of Native Biodiversity on Islands', *PNAS* 105, supplement 1 (12 August 2008): 11490–97, doi:10.1073/pnas.0802290105
27. Holmes, 'Day of the Sparrow'.
28. Mark Davis, 'Biotic Globalization: Does Competition from Introduced Species Threaten Biodiversity?', *BioScience* 53, no. 5 (2003): 481–89, doi:10.1641/0006-3568(2003)053[0481:BGDCFI]2.0.CO;2
29. Davis, 'Researching Invasive Species 50 Years After Elton'.
30. R. Travis Belote et al., 'Diversity-Invasibility Across an Experimental Disturbance Gradient in Appalachian Forests', *Ecology* 89, no. 1 (January 2008): 183–92, doi:10.1890/07-0270.1
31. Daniel Simberloff et al., 'Impacts of Biological Invasions: What's What and the Way Forward', *Trends in Ecology and Evolution* 28, no. 1 (January 2013): 58–66, doi:10.1016/j.tree.2012.07.013
32. David M. Richardson and Anthony Ricciardi, 'Misleading Criticisms of Invasion Science: A Field Guide', *Diversity and Distributions* 19 (2013), 1461–67.
33. Benjamin Gilbert and Jonathan Levine, 'Plant Invasions and Extinction Debts', *PNAS* 110 (2013): 1744–49, http://www.ncbi.nlm.nih.gov/pmc/articles/PMC3562839
34. Stephen T. Jackson and Dov F. Sax, 'Balancing Biodiversity in a Changing Environment: Extinction Debt, Immigration Credit and

Species Turnover', *Trends in Ecology and Evolution* 25, no. 3 (2010): 153–60, doi:10.1016/j.tree.2009.10.001
35. 'Impact of Invasive Alien Species', WWF, http://wwf.panda.org/about_our_earth/species/problems/invasive_species/, accessed December 2013.
36. Matthew Chew and Andrew Hamilton, 'The Rise and Fall of Biotic Nativeness: A Historical Perspective', in *Fifty Years of Invasion Ecology*, ed. David Richardson, 35–47.
37. Esteban Paolucci et al., 'Origin Matters: Alien Consumers Inflict Greater Damage on Prey Populations than do Native Consumers', *Diversity and Distributions* 19 (January 2013): 988–95, doi:10.1111/ddi.12073
38. Daniel Simberloff et al., 'The Natives Are Restless, but Not Often and Mostly When Disturbed', *Ecology* 93, no. 3 (March 2012): 598–607, doi:10.1890/11-1232.1
39. Lawrence Slobodkin, 'The Good, the Bad and the Reified', *Evolutionary Ecology Research* 3 (2001): 1–13.
40. Achim Steiner, 'Counting the Cost of Alien Invasions', BBC News Online, 13 April 2010, http://news.bbc.co.uk/1/hi/sci/tech/8615398.stm
41. Pimentel et al., 'Economic and Environmental Threats of Alien Plant, Animal, and Microbe Invasions'.
42. Ibid.
43. Pimentel et al., 'Update on the Environmental and Economic Costs Associated with Alien-Invasive Species'.
44. Scott Loss et al., 'The Impact of Free-Ranging Domestic Cats on Wildlife of the United States', *Nature Communications* 4 (2013): 1396, doi:10.1038/ncomms2380
45. Ellen Bjerkås, *The Economic Importance of Companion Animals* (Brussels: Federation of European Companion Animal Veterinary Associations, 2007), http://www.fecava.org/
46. Pimentel et al., 'Update on the Environmental and Economic Costs Associated with Alien-Invasive Species'.
47. Mark Davis, 'Invasion Biology, 1958–2005: The Pursuit of Science and Conservation', in *Conceptual Ecology and Invasion Biology: Reciprocal Approaches to Nature*, Invading Nature series, ed. Marc W. Cadotte et al. (Dordrecht, Netherlands: Springer, 2006), 35–64.
48. Loïc Valéry et al., 'Another Call for the End of Invasion Biology', *Oikos* 122, no. 8 (August 2013): 1143–46, doi:10.1111/j.1600-0706.2013.00445.x

49. Marco Lambertini et al., 'Invasives: A Major Conservation Threat', *Science* 333, no. 6041 (July 2011): 404–5, doi:10.1126/science.333.6041.404-b
50. Lesley M. Head and Pat Muir, 'Nativeness, Invasiveness and Nation in Australian Plants', University of Wollongong Research Online, 2004, http://ro.uow.edu.au/cgi/viewcontent.cgi?article=1084&context=scipapers

Chapter 8: Myths of the Pristine

1. Cited in Jose Toribio Medina and Bertram T Lee, The Discovery of the Amazon According to the Account of Friar Gaspar de Carvajal and Other Documents, American Geographical Society, 1934; 217
2. John Noble Wilford, 'Sharp and to the Point in Amazonia', *New York Times*, 23 April 1997, http://www.nytimes.com/1997/04/23/science/sharp-and-to-the-point-in-amazonia.html
3. Michael Heckenberger et al., 'Amazonia 1492: Pristine Forest or Cultural Parkland?', *Science* 301, no. 5640 (19 September 2003): 1710–14, doi:10.1126/science.1086112
4. Doyle McKey et al., 'Pre-Columbian Agricultural Landscapes, Ecosystem Engineers, and Self-Organized Patchiness in Amazonia', *PNAS* 107, no. 17 (2010): 7823–28, doi:10.1073/pnas.0908925107
5. Michael Fay and Steve Blake, *Evidence of Secondarization in Dense Forest of Northern Congo and Southwestern Central African Republic Between 2340 and 990 B.P.* (New York: Wildlife Conservation Society, 1998).
6. Germain Bayon et al., 'Intensifying Weathering and Land Use in Iron Age Central Africa', *Science* 335, no. 6073 (March 2012): 1219–22, doi:10.1126/science.1215400
7. Richard Oslisly et al., 'Climatic and Cultural Changes in the West Congo Basin Forests over the Past 5000 Years', *Philosophical Transactions of the Royal Society B* 368, no. 1625 (23 July 2013): 201, doi:10.1098/rstb.2012.0304
8. Lee J.T. White and John F. Oates, 'New Data on the History of the Plateau Forest of Okomu, Southern Nigeria: An Insight into How Human Disturbance Has Shaped the African Rain Forest', *Global Ecology and Biogeography* 8, no. 5 (September 1999): 355–61, doi:10.1046/j.1365-2699.1999.00149.x
9. Kathy J. Willis et al., 'How "Virgin" Is Virgin Rainforest?', *Science* 304, no. 5669 (16 April 2004): 402–3, doi:10.1126/science.1093991

10. Barend van Gemerden et al., 'The Pristine Rain Forest? Remnants of Historical Human Impacts on Current Tree Species Composition and Diversity', *Journal of Biogeography* 30 (2003): 1381–90.
11. Christopher Hunt and Ryan Rabett, 'Holocene Landscape Intervention and Plant Food Production Strategies in Island and Mainland Southeast Asia', *Journal of Archaeological Science*, in press 2013, doi:10.1016/j.jas.2013.12.011
12. Fred Pearce, 'A Gift from the Mayans', *New Scientist*, 26 May 2006, 48–9.
13. Carol Kaesuk Yoon, 'Rain Forests Seen as Shaped by Human Hand', *New York Times*, 27 July 1993, http://www.nytimes.com/1993/07/27/science/rain-forests-seen-as-shaped-by-human-hand.html
14. Charles Mann, 'The Real Dirt on Rainforest Fertility', *Science* 297, no. 5583 (9 August 2002): 920–22, doi:10.1126/science.297.5583.920
15. Melissa Leach et al., 'Green Grabs and Biochar: Revaluing African Soils and Farming in the New Carbon Economy', *Journal of Peasant Studies* 39, no. 2 (2012): 285–307, doi:10.1080/03066150.2012.658042
16. Douglas Sheil et al., 'Do Anthropogenic Dark Earths Occur in the Interior of Borneo? Some Initial Observations from East Kalimantan', *Forests* 3, no. 2 (2012): 207–29, doi:10.3390/f3020207
17. American Geophysical Union, 'Ancient Trash Heaps Gave Rise to Everglades Tree Islands', press release, 21 March 2011, http://news.agu.org/press-release/ancient-trash-heaps-gave-rise-to-everglades-tree-islands/
18. Colin Tudge, *Neanderthals, Bandits and Farmers: How Agriculture Really Began* (New Haven, CT: Yale University Press, 1999).
19. Amy Bogaard et al., 'Crop Manuring and Intensive Land Management by Europe's First Farmers', *PNAS* 110, no. 31 (30 July 2013): 12589–94, doi:10.1073/pnas.1305918110
20. David M.J.S. Bowman et al., 'The Human Dimension of Fire Regimes on Earth', *Journal of Biogeography* 38, no. 12 (December 2011): 2223–36, doi:10.1111/j.1365-2699.2011.02595.x
21. Erle C. Ellis et al., 'Used Planet: A Global History', *PNAS* 110, no. 20 (2013): 7978–85, doi:10.1073/pnas.1217241110
22. David Briggs, *Plant Microevolution and Conservation in Human-influenced Ecosystems* (New York: Cambridge University Press, 2009), 86.

23. Oliver Rackham, 'Historic Land-Use Patterns', in *Encyclopedia of Biodiversity, Volume 3*, ed. Simon Levin (Waltham, MA: Academic Press, 2001), 675–87.
24. Willis et al., 'How "Virgin" Is Virgin Rainforest?'.
25. Theodore Roosevelt, *African Game Trails: The Classic Big Game Safari* (Torrington, WY: Narrative Press, 2001).
26. John Reader, *Africa: A Biography of the Continent* (London: Hamish Hamilton, 1997).
27. Frederick Lugard, *The Rise of Our East African Empire* (Edinburgh: W. Blackwood and Sons, 1893), 525.
28. Ibid., 526.
29. Robin Reid, *Savannas of Our Birth: People, Wildlife, and Change in East Africa* (Berkeley: University of California Press, 2012), 108.
30. Reader, *Africa*.
31. Lugard, *The Rise of Our East African Empire*, 527.
32. 'Huxley in Africa', *Observer* (UK), 13 November 1960.
33. Bernhard Grzimek and Michael Grzimek, *Serengeti Shall Not Die* (London: Hamish Hamilton, 1960).
34. Douglas Sheil and Erik Meijaard, 'Purity and Prejudice: Deluding Ourselves About Biodiversity Conservation', *Biotropica* 42(5), 2010, 566–8, doi: 10.1111/j.1744-7429.2010.00687.x

Chapter 9: Nativism in the Garden of Eden

1. 'George Perkins Marsh: Renaissance Vermonter', George Perkins Marsh Institute, Clark University, http://www.clarku.edu/departments/marsh/about/, accessed January 2014.
2. George Perkins Marsh, *Man and Nature; or, Physical Geography as Modified by Human Action* (1864; reprinted Seattle: University of Washington Press, 2003), 28–30, 42–3, 264.
3. Stephen Trudgill, 'A Requiem for the British Flora? Emotional Biogeographies and Environmental Change', *Area* 40, no. 1 (March 2008): 99–107, doi:10.1111/j.1475-4762.2008.00790.x
4. Daniel Botkin, *The Moon in the Nautilus Shell: Discordant Harmonies Reconsidered* (New York: Oxford University Press, 2012).
5. Frederic E. Clements, *Plant Succession: An Analysis of the Development of Vegetation* (Washington, DC: Carnegie Institution of Washington, 1916).
6. Dave Hone, 'Moth Tongues, Orchids and Darwin – the Predictive Power of Evolution', *Guardian Online* (UK), 2 October 2013,

http://www.theguardian.com/science/lost-worlds/2013/oct/02/moth-tongues-orchids-darwin-evolution

7. 'Jean-Baptiste Lamarck', *Human Evolution* (Andrew Lehman website), http://serpentfd.org/b/lamarck.html, accessed January 2014.
8. James Lovelock, *Gaia: A New Look at Life on Earth* (Oxford, UK: Oxford University Press, 1979).
9. Henry Gleason, 'The Individualistic Concept of the Plant Association', *Bulletin of the Torrey Botanical Club* 53 (1926): 7–26, http://www.ecologia.unam.mx/laboratorios/comunidades/pdf/pdf%20curso%20posgrado%20Elena/Tema%201/gleason1926.pdf
10. Gleason and Barbour quotes: Michael Barbour, 'Ecological Fragmentation of the Fifties', in *Uncommon Ground: Rethinking the Human Place in Nature*, ed. William Cronon (New York: W.W. Norton, 1995), 238.
11. Eugene P. Odum, *Fundamentals of Ecology* (Philadelphia: W.B. Saunders, 1953).
12. Eugene P. Odum, 'The Strategy of Ecosystem Development', *Science* 164, no. 3877 (18 April 1969): 262–70, doi:10.1126/science.164.3877.262
13. Paul Ehrlich et al., 'No Middle Way on the Environment', *Atlantic Monthly*, December 1997, 98–104, http://www.theatlantic.com/past/docs/issues/97dec/enviro.htm
14. Robert M. May, 'Simple Mathematical Models with Very Complicated Dynamics', *Nature* 261 (10 June 1976), 459–67, doi:10.1038/261459a0
15. Robert May, 'The Chaotic Rhythms of Life', *New Scientist*, 18 November 1989, http://www.newscientist.com/article/mg12416913.500
16. Stephen Jay Gould, 'An Evolutionary Perspective on Strengths, Fallacies, and Confusions in the Concept of Native Plants', *Arnoldia* (Spring 1998), http://arnoldia.arboretum.harvard.edu/pdf/articles/483.pdf
17. Ibid.
18. Ibid.
19. Mark A. Davis et al., 'Don't Judge Species on Their Origins', *Nature* 474 (9 June 2011): 153–4, doi:10.1038/474153a
20. William Cronon, 'Modes of Prophecy and Production: Placing Nature in History', *Journal of American History* 76, no. 4 (March 1990): 1122–31.

21. Daniel H. Janzen, '*Enterolobium cyclocarpum* Seed Passage Rate and Survival in Horses, Costa Rican Pleistocene Seed Dispersal Agents', *Ecology* 62, no. 3 (June 1981): 593–601, http://www.jstor.org/stable/1937726
22. Daniel H. Janzen, 'When Is It Co-evolution?', *Evolution* 34, no. 3 (May 1980): 611–12.
23. Ibid.
24. Fred Pearce, 'Wild About Fire', *New Scientist*, 11 November 2000, 50–51.
25. Bob Holmes, 'Survival of the Weakest', *New Scientist*, 30 January 1999, 15.
26. Stephen P. Hubbell, *The Unified Neutral Theory of Biodiversity and Biogeography* (Princeton, NJ: Princeton University Press, 2001).
27. Hans ter Steege et al., 'Hyperdominance in the Amazonian Tree Flora', *Science* 342 (2013): 325–35, doi:10.1126/science.1243092
28. Botkin, *The Moon in the Nautilus Shell.*
29. 'Nomination of Surtsey for the UNESCO World Heritage List', UNESCO, 2007, http://whc.unesco.org/uploads/nominations/1267.pdf
30. 'Surtsey', Surtsey Research Society, http://www.surtsey.is/pp_ens/biola_1.htm, accessed September 2013.
31. Borgthor Magnusson et al., 'Developments in Plant Colonization and Succession on Surtsey During 1999–2008', *Surtsey Research* 12 (2009): 57–76, http://www.surtsey.is
32. Ian Thornton, *Island Colonization: The Origin and Development of Island Communities* (Cambridge, UK: Cambridge University Press, 2007).
33. Charles Elton, *The Ecology of Invasions by Animals and Plants* (1958; reprinted, Chicago: University of Chicago Press, 2000).
34. Thornton, *Island Colonization.*

Chapter 10: Novel Ecosystems

1. Thomas K. Rudel et al., 'When Fields Revert to Forests: Development and Spontaneous Reforestation in Post-War Puerto Rico', *Professional Geographer* 52, no. 3 (2000): 386–97, doi:10.1111/0033-0124.00233
2. Ariel E. Lugo and Eileen Helmer, 'Emerging Forests on Abandoned Land: Puerto Rico's New Forests', *Forest Ecology and Management* 190 (January 2004): 145–61, doi:10.1016/j.foreco.2003.09.012

3. Oscar Abelleira Martinez, 'Observations on the Fauna that Visit African Tulip Tree (*Spathodea campanulata Beauv.*) Forests in Puerto Rico', *Acta Cientifica* 22, nos. 1–3 (2008): 37–48; available online: Treesearch, US Forest Service, http://www.treesearch.fs.fed.us/pubs/38141
4. Ariel E. Lugo et al., 'Natural Mixing of Species: Novel Plant–Animal Communities on Caribbean Islands', *Animal Conservation* 15, no. 3 (2012): 233–41, doi:10.1111/j.1469-1795.2012.00523.x
5. Gaia Vince, 'Embracing Invasives', *Science* 331, no. 6023 (18 March 2011): 1383–84, doi:10.1126/science.331.6023.1383
6. Lugo et al., 'Natural Mixing of Species'.
7. '*Eleutherodactylus coqui*', IUCN Red List of Threatened Species, http://www.iucnredlist.org/details/56522/0, accessed December 2013.
8. Daniel Botkin, *The Moon in the Nautilus Shell: Discordant Harmonies Reconsidered* (New York: Oxford University Press, 2012).
9. Fred Pearce, 'Deforestation in Congo Basin Slows, but for How Long?', *New Scientist online*, 24 July 2013, http://www.newscientist.com/article/dn23924
10. Francis E. Putz et al., 'Sustaining Conservation Values in Selectively Logged Tropical Forests: The Attained and the Attainable', *Conservation Letters* 5, no. 4 (August 2012): 296–303, doi:10.1111/j.1755-263X.2012.00242.x
11. Kathy J. Willis et al., 'How "Virgin" Is Virgin Rainforest?', *Science* 304, no. 5669 (16 April 2004): 402–3, doi:10.1126/science.1093991
12. Ibid.
13. Carnegie Institution, '80 Percent of Malaysian Borneo Degraded by Logging', Science Daily, press release, 17 July 2013, http://www.sciencedaily.com/releases/2013/07/130717173002.htm
14. Jane E. Bryan et al., 'Extreme Differences in Forest Degradation in Borneo: Comparing Practices in Sarawak, Sabah, and Brunei', *PLoS One* 8, no. 7 (17 July 2013), doi:10.1371/journal.pone.0069679
15. Fred Pearce, 'Hit and Run in Sarawak', *New Scientist*, 12 May 1990, 46–9.
16. David Edwards et al., 'Degraded Lands Worth Protecting: The Biological Importance of Southeast Asia's Repeatedly Logged Forests', *Proceedings of the Royal Society B* 278 (2011): 82–90, rspb.2010.1062v1

17. Daisy H. Dent, 'Defining the Conservation Value of Secondary Tropical Forests', *Animal Conservation* 13, no. 1 (January 2010): 14–15, doi:10.1111/j.1469-1795.2010.00346.x
18. Robin L. Chazdon et al., 'The Potential for Species Conservation in Tropical Secondary Forests', *Conservation Biology* 23, no. 6 (2009): 1406–17, doi:10.1111/J.1523-1739.2009.01338.X
19. Richard J. Hobbs et al., 'Novel Ecosystems: Theoretical and Management Aspects of the New Ecological World Order', *Global Ecology and Biogeography* 15, no. 1 (January 2006): 1–7, doi:10.1111/j.1466-822X.2006.00212.x
20. Charles Elton, *The Ecology of Invasions by Animals and Plants* (1958; reprinted, Chicago: University of Chicago Press, 2000).
21. Stephen M. Meyer, *The End of the Wild* (Cambridge, MA: MIT Press, 2006).
22. Sharon Levy, 'As Larger Animals Decline, Forests Feel Their Absence', *Yale Environment 360*, 31 March 2011, http://e360.yale.edu/feature/as_larger_animals_decline_forests_feel_their_absence/2381, accessed September 2014.
23. Christoff Kueffer et al., 'Case Study: Management of Novel Ecosystems in the Seychelles', in *Novel Ecosystems: Intervening in the New Ecological World Order*, ed. Richard Hobbs et al. (Hoboken, NJ: Wiley-Blackwell, 2013), 228–38.
24. Vince, 'Embracing Invasives'.
25. Segundo Coello and Alan Saunders, *Final Project Evaluation: Control of Invasive Species in the Galapagos Archipelago, ECU/00/G31*, prepared for the Global Environment Fund (GEF)/United Nations Development Programme (UNDP)/[Ecuador] Ministry of the Environment (MAE), 2011, http://www.thegef.org/gef/sites/thegef.org/files/gef_prj_docs/GEFProjectDocuments/M&E/TER/FY2011/UNDP/763/1349_BD_Ecuador_TE.pdf
26. Mark Gardener, 'Implementing Novel Ecosystem Management in the Galapagos Islands', Charles Darwin Foundation, Galapagos, Ecuador, http://www.restorationinstitute.ca/wp-content/uploads/2011/10/Gardener-reduced.pdf
27. Daniel Simberloff, *Invasive Species: What Everyone Needs to Know* (New York: Oxford University Press, 2013).
28. Victor Carrion et al., 'Archipelago-Wide Island Restoration in the Galápagos Islands: Reducing Costs of Invasive Mammal Eradication

Programs and Reinvasion Risk', *PLoS One* 6, no. 5 (11 May 2011), doi:10.1371/journal.pone.0018835

29. Holly Jones and Oswald Schmitz, 'Rapid Recovery of Damaged Ecosystems', *PLoS One* 4, no. 5 (2009): e5653, doi:10.1371/journal.pone.0005653
30. *Ecosystems and Human Well-Being*, Millennium Ecosystem Assessment, 2005, http://www.maweb.org/en/index.aspx
31. Andrew Light et al., 'Valuing Novel Ecosystems', in *Novel Ecosystems*, 257–68.

Chapter 11: Rebooting Conservation in the Urban Badlands

1. Euan Stretch, 'Rare Jumping Spiders Delay Building Work on £2bn Paramount Theme Park in Kent', *Daily Mirror* (UK), 4 April 2013, http://www.mirror.co.uk/news/uk-news/spiders-delay-2bn-paramount-theme-1811793
2. 'Canvey Island – Britain's Rainforest', Buglife, http://www.buglife.org.uk/campaigns-and-our-work/habitat-projects/canvey-islan d-britains-rainforest, accessed October 2013.
3. John Vidal, 'A Bleak Corner of Essex is Being Hailed as England's Rainforest', *Guardian* (UK), 3 May 2003, http://www.theguardian.com/uk/2003/may/03/ruralaffairs.science
4. Peter Shaw and Wes Halton, 'Classic Sites: Nob End, Bolton', *British Wildlife*, October 1998, 13–17.
5. P.R. Harvey, 'Brownfield Importance to Invertebrates', Essex Field Club, 2006, http://www.essexfieldclub.org.uk/portal/p/Brownfield+importance
6. 'Proposal for SSSI – Chattenden Woods and Lodge Hill, Kent', Natural England Executive Board, meeting no. 88, 11 March 2013, http://www.naturalengland.org.uk/Images/lodge-hill-meeting-paper-11march2012_tcm6–35509.pdf
7. D. McKlintock, 'J.E. Lousley and Plants Alien in the British Isles', *Watsonia* 11 (1977): 287–90, http://archive.bsbi.org.uk/Wats11p287.pdf
8. Richard Fitter and Edward Lousley, *The Natural History of the City* (London: Corporation of the City, 1953).
9. D. McKlintock, 'J.E. Lousley and Plants Alien in the British Isles'.
10. Edward Lousley, 'How Sheep Influence the Travel of Plants', *New Scientist* 8 (1960): 353–4.
11. Peter Marren, 'Oliver Gilbert', obituary, *Independent* (UK),

18 May 2005, http://www.independent.co.uk/news/obituaries/oliver-gilbert-6145986.html

12. Oliver Gilbert, *The Flowering of the Cities: The Natural Flora of 'Urban Commons'* (Peterborough, UK: English Nature, 1992).
13. Michael Crawley et al., 'The Population Biology of Invaders', *Philosophical Transactions of the Royal Society B* 314 (1986): 711–31.
14. Michael P. Perring et al., 'Incorporating Novelty and Novel Ecosystems into Restoration Planning and Practice in the 21st Century', *Ecological Processes* 2, no. 18 (2013), doi:10.1186/2192-1709-2-18
15. 'Urban Jungle Chichester', YouTube, http://www.youtube.com/watch?v=xf6t-C7y52o&feature=youtu.be, accessed February 2014.
16. Roberta Kruger, 'PBS Documentary *Raccoon Nation* Suggests Humans Are Making Raccoons Smarter', *Treehugger*, 7 February 2012, http://www.treehugger.com/culture/pbs-documentary-raccoon-nation-humans-making-raccoons-smarter.html
17. Sarah DeWeerdt, 'Cohabitation', *Conservation*, 11 December 2013, http://conservationmagazine.org/2013/12/cohabitation
18. 'Dogged Persistence', *Economist*, 9 March 2013, http://www.economist.com/news/united-states/21573167-coyote-quietly-conquering-urban-america-dogged-persistence
19. James Barilla, 'Accidental Conservation', *Conservation*, 11 December 2013, http://conservationmagazine.org/2013/12/accidental-conservation
20. Michael McCarthy, 'The Secret Life of Sparrows', *Independent* (UK), 2 August 2006.
21. Humphrey Crick et al., 'Investigation into the Causes of the Decline of Starlings and House Sparrows in Great Britain', (Thetford, UK: British Trust for Ornithology, 2002), http://www.bto.org/sites/default/files/u32/researchreports/rr290.pdf
22. 'Birds Sing Louder Amidst the Noise and Structures of the Urban Jungle', press release, University of Copenhagen, 22 February 2012, http://news.ku.dk/all_news/2012/2012.2/birds-sing-louder-amidst-the-noise-and-structures-of-the-urban-jungle/
23. Roger Highfield, 'Pigeons Hop on the Tube to Save Their Wings', *Daily Telegraph* (UK), 25 March 1996.
24. 'About', Nature of Cities, http://www.thenatureofcities.com/about, accessed February 2014.
25. 'What Can We Do to Improve Pollinator Diversity and Abundance

in Urban Areas?', Research Questions, Urban Pollinators Project, University of Bristol, http://www.bristol.ac.uk/biology/research/ecological/community/pollinators/question3, accessed February 2014.

26. 'Biodiversity and Natura2000 in Urban Areas: A Review of Key Issues and Experiences in Europe', European Urban Knowledge Network, http://www.eukn.org/E_library/Urban_Environment/Environmental_Sustainability/Environmental_Sustainability/Biodiversity_and_Natura2000_in_urban_areas_a_review_of_key_issues_and_experiences_in_Europe, accessed February 2014.
27. 'Invasive Alien Species: The Urban Dimension', IUCN, http://www.iucn.org/news_homepage/news_by_date/?13588/Invasive-alien-species-the-urban-dimension, accessed February 2014.
28. Mark Kinver, 'Invasive Alien Species Threaten Urban Environments', BBC News, 5 September 2013, http://www.bbc.co.uk/news/science-environment-23936489
29. Alison Loram et al., 'Urban Domestic Gardens (XII): The Richness and Composition of the Flora in Five UK cities', *Journal of Vegetation Science* 19, no. 3 (2008): 321–30.
30. *State of the Natural Environment 2008* (Sheffield, UK: Natural England, 2008), http://www.naturalengland.org.uk/
31. Adrian Spalding, *The Butterfly Handbook: General Advice Note on Mitigating the Impact of Roads on Butterfly Populations* (Sheffield, UK: English Nature, 2005), http://publications.naturalengland.org.uk/publication/130003
32. Jim Smith and Dave Timms, 'Nuclear Energy Greener Than You Think?', *Planet Earth*, Spring 2006, 22–4.
33. Heather Meeks et al., 'Understanding the Genetic Consequences of Environmental Toxicant Exposure: Chernobyl as a Model System', *Environmental Toxicology and Chemistry* 28, no. 9 (2009): 1982–94.
34. 'Radiation-Ecological Monitoring', Polesye State Radiation Ecological Reserve, http://www.zapovednik.by/en/research/monitoring/, accessed February 2014.
35. 'Coral Flourishing at Bikini Atoll Atomic Test Site', NBCNews.com, 15 April 2008, http://www.nbcnews.com/id/24132798
36. Peter Kareiva et al., 'Conservation in the Anthropocene', *The Breakthrough*, Winter 2012, http://thebreakthrough.org/index.php/journal/past-issues/issue-2/conservation-in-the-anthropocene

37. Ibid.
38. Ibid.
39. 'Rebuilding Not Rewinding Is the Future of Conservation', *New Scientist*, 16 May 2013, 3.

Chapter 12: Call of the New Wild

1. Chris D. Thomas et al., 'Extinction Risk from Climate Change', *Nature* 427 (8 January 2004): 145–8, doi:10.1038/nature02121
2. University of Durham, 'Assisted Colonisation Key to Species' Survival in Changing Climate', press release, 18 February 2009, https://www.dur.ac.uk/news/newsitem/?itemno=7606
3. Anthony Ricciardi and Daniel Simberloff, 'Assisted Colonization is Not a Viable Conservation Strategy', *Trends in Ecology and Evolution* 24 (2009): 248–53, doi:10.1016/j.tree.2008.12.006
4. Chris Thomas, 'Britain Should Welcome Climate Refugee Species', *New Scientist*, 2 November 2011, 29–30.
5. Ibid.
6. Chris Thomas, 'The Anthropocene Could Raise Biological Diversity', *Nature* 502, no. 7 (2 October 2013): 7, doi:10.1038/502007a
7. Quoted in Carl Zimmer, 'First Comes Global Warming, Then an Evolutionary Explosion', *Yale Environment 360*, 3 August 2009, http://e360.yale.edu/feature/first_comes_global_warming_then_an_evolutionary_explosion/2178, accessed September 2014.
8. Tom C. Cameron et al., 'Eco-evolutionary Dynamics in Response to Selection on Life-History', *Ecology Letters* 16, no. 6 (June 2013): 754–63, doi:10.1111/ele.12107
9. Ibid.
10. Martin Wainwright, 'Blooming Unexpected', *Guardian* (UK), 20 February 2003, http://www.theguardian.com/uk/2003/feb/20/science.highereducation
11. Richard F. Johnston and Robert K. Selander, 'House Sparrows: Rapid Evolution of Races in North America', *Science* 144 (1964): 548–50.
12. Andrew P. Hendry et al., 'Rapid Evolution of Reproductive Isolation in the Wild: Evidence from Introduced Salmon', *Science* 290, no. 516 (2000): 516–18, doi:10.1126/science.290.5491.516
13. Dietmar Schwarz et al., 'Host Shift to an Invasive Plant Triggers Rapid Animal Hybrid Speciation', *Nature* 436 (28 July 2005): 546–9, doi:10.1038/nature03800

14. Daniel Simberloff, *Invasive Species: What Everyone Needs to Know* (New York: Oxford University Press, 2013).
15. Fred Pearce, 'Human Meddling Will Spur On the Evolution of New Species', *New Scientist*, 17 January 2014, 28–9.
16. Christoph Kueffer and Christopher Kaiser-Bunbury, 'Reconciling Conflicting Perspectives for Biodiversity Conservation in the Anthropocene', *Frontiers in Ecology and the Environment* 12, no. 2 (March 2014): 131–7, doi:10.1890/120201
17. Martin A. Schlaepfer et al., 'The Potential Conservation Value of Non-Native Species', *Conservation Biology* 25, no. 3 (June 2011): 428–37, doi:10.1111/j.1523-1739.2010.01646.x
18. Daniel Botkin, *The Moon in the Nautilus Shell: Discordant Harmonies Reconsidered* (New York: Oxford University Press, 2012).
19. Stephen Jackson, 'Perspective: Ecological Novelty Is Not New', in *Novel Ecosystems: Intervening in the New Ecological World Order*, ed. Richard Hobbs et al. (Hoboken, NJ: Wiley-Blackwell, 2013), 63–5.
20. Stephen M. Meyer, *The End of the Wild* (Cambridge, MA: MIT Press, 2006).
21. Garry Hamilton, *Super Species: The Creatures That Will Dominate the Planet* (Richmond Hill, Ontario, Canada: Firefly, 2010).
22. Kathy H. Hodder et al., 'Can the Pre-Neolithic Provide Suitable Models for Re-wilding the Landscape in Britain?', *British Wildlife* 20, no. 5, special supplement (June 2009): 4–15, http://www.britishwildlife.com/
23. Eric Bignal and Davy McCracken, 'Herbivores in Space: Extensive Grazing Systems in Europe', *British Wildlife* 20, no. 5, special supplement (June 2009): 44–9.
24. Alladale Wilderness Reserve, http://www.alladale.com/, accessed January 2014.
25. Frans W.M. Vera, 'Large-Scale Nature Development – the Oostvaardersplassen', *British Wildlife* 20, no. 5, special supplement (June 2009): 28–36.
26. American Prairie Reserve, http://www.americanprairie.org/, accessed January 2014.
27. Ted's Montana Grill, http://www.tedsmontanagrill.com/, accessed January 2014.
28. Pleistocene Park, http://www.pleistocenepark.ru/en/, accessed January 2014.

29. Stewart Brand, 'The Case for De-Extinction: Why We Should Bring Back the Woolly Mammoth', *Yale Environment 360*, 13 January 2014, http://e360.yale.edu/feature/the_case_for_de-extinction_why_we_should_bring_back_the_woolly_mammoth/2721
30. 'Professor Arthur Shapiro Comments on the Environmental Impact Report of the Natural Areas Program in San Francisco', *Coyote Yips*, 19 October 2011, http://coyoteyipps.com/2011/10/19/professor-arthur-shapiro-comments-on-the-environmental-impact-report-of-the-natural-areas-program-in-san-francisco, accessed January 2014.
31. Daniel Simberloff and Donald Strong, 'Counterpoint: Scientists Offer a Dissenting View on Ascension Island', *Yale Environment 360*, 2013, http://e360.yale.edu/Counterpoint_Scientists_Offer_Dissenting_View_on_Ascension_Island.msp

Index